STUDENT'S SOLUTIONS MANUAL

JONES/CHILDERS

CONTEMPORARY COLLEGE PHYSICS

2ND EDITION

EDWIN JONES
THE UNIVERSITY OF SOUTH CAROLINA

RICHARD CHILDERS
THE UNIVERSITY OF SOUTH CAROLINA

Addison-Wesley Publishing Company

Reading, Massachusetts • Menlo Park, California • New York
Don Mills, Ontario • Wokingham, England • Amsterdam • Bonn
Sydney • Singapore • Tokyo • Madrid • San Juan • Milan • Paris

The authors gratefully acknowledge the work of Prof. James Whitenton, Prof. Fred Watts, Prof. Clyde Smith, and Mr. Craig Watkins in working the problems in the text.

Reproduced by Addison-Wesley from camera-ready copy supplied by the authors.

ISBN 0-201-62960-7

5 6 7 8 9 10-BA-9594

TABLE OF CONTENTS

SOLUTIONS TO THE ODD-NUMBERED PROBLEMS

Chapter 1

MEASUREMENT AND ANALYSIS

1.1 First draw a diagram of the situation. The diagonal is obtained from the Pythagorean theorem as $L = \sqrt{(50.0\text{ m})^2 + (100\text{ m})^2} = 112$ m. One fourth of the diagonal is (112 m)/4 = 28.0 m.

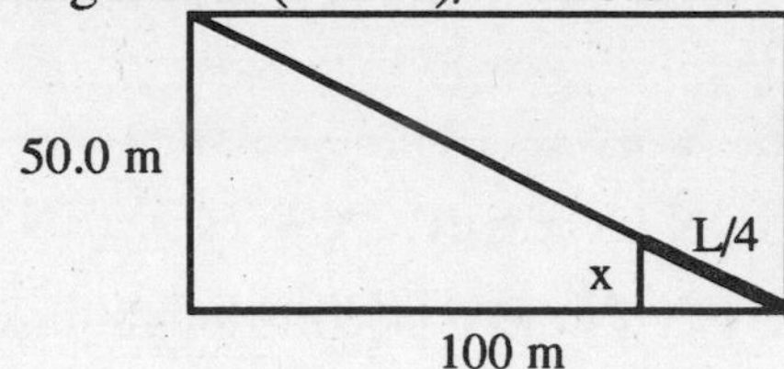

The length to the nearest side x is found by similar triangles:

$$\frac{x}{28.0\text{ m}} = \frac{50.0\text{ m}}{112\text{ m}}. \quad \boxed{x = 12.5\text{ m.}}$$

1.3 From the definition of an angle: $s = r\theta$. Here r = 7.5 cm and

$$\theta = 1° = 1° \left(\frac{2\pi\text{ rad}}{360°}\right) = \frac{2\pi\text{ rad}}{360}.$$

$s = 7.5\text{ cm} \times \frac{2\pi}{360}\text{ rad} = 0.13$ cm, or $\boxed{s = 1.3\text{ mm}}$. (Remember that the radian is dimensionless.)

1.5 We rewrite the defining equation for an angle in radian measure as $s = r\theta$. Then $s = (100\text{ m})(0.020\text{ rad}) = \boxed{2.0\text{ m}}$.

1.7 Convert the rate of rotation of the moon from units of degrees per day to units of degrees per hour. Then the rotation in any time interval is the product of the rate with the time which in this case is 1 h. So $\frac{360°}{27.3\text{ days}} \times \frac{1\text{ day}}{24\text{ h}} \times 1\text{ h} = \boxed{0.549°}$.

1.9 Make your own graph and count the squares. Remember that the area swept out will be the same for both motions, so the number of squares should be the same.

1.11 Kepler's third law (Eq. 1.2) may be rearranged to get $R = \left(\frac{T^2}{k}\right)^{1/3}$. Here we use $T = 29.46$ year and $k = 1\text{ year}^2/\text{AU}^3$ to get $\boxed{R = 9.54\text{ AU}}$. This answer can also be given in meters as $R = 9.54\text{ AU} \times (1.5 \times 10^{11}\text{ m/AU}) = 1.4 \times 10^{12}$ m.

1.13 Kepler's third law holds for objects orbiting the earth if the proper k value is used. $k_e = \frac{T^2}{R^3} = \frac{(T_{moon})^2}{(R_{moon})^3}$, where $T_{moon} = 27.33$ days and $R_{moon} = 3.84 \times 10^8$ m. The radius of the satellite orbit is then obtained from

$$R = \left(\frac{T^2}{k_e}\right)^{1/3} = \left(\frac{(1\ \text{day})^2}{(27.33\ \text{d})^2/(3.84 \times 10^8\ \text{m})^3}\right)^{1/3} = \boxed{4.23 \times 10^7\ \text{m}}.$$

1.15 The second hand makes a full revolution every minute. In three years it goes around once for each minute elapsed. In 3 years it goes around by 3 years $\times$ 365 day/year $\times$ 24 h/day $\times$ 60 min/h = 1.58×10^6 min. So in 3 years the second hand makes $\boxed{1.58 \times 10^6 \text{ revolutions}}$.

1.17 Volume = length $\times$ width $\times$ height
volume = (115 cm)(31.4 cm)(6.5 cm) = 23471.5 cm^3. Since the height is known to only 2 significant figure we must round off the answer to. $\boxed{2.3 \times 10^4\ \text{cm}^3}$.

1.19 One year = 365 day $\times$ 24 h/day $\times$ 60 min/h $\times$ 60 s/min = $\boxed{3.2 \times 10^7\ \text{s}}$.

1.21 Population density = population/land area.

$$\text{Population density} = \frac{2.5 \times 10^8\ \text{people}}{9.4 \times 10^6\ \text{km}^2} = \boxed{27\ \text{people/km}^2}.$$

1.23 The total price is the number of sheets times the price per sheet.
Total price = 12×10^6 sheets $\times$ 0.18 cents/sheet = 2.16×10^6 cents.
In dollars we get 2.16×10^6 cents $\times$ 10^{-2} dollars/cent = $\boxed{\$\ 2.16 \times 10^4}$.
This can also be expressed as \$ 21,600.

1.25 Time for light to travel from the sun = distance to the sun/speed of light.

$$t = \frac{1.50 \times 10^{11}\ \text{m}}{3.00 \times 10^8\ \text{m/s}} = \boxed{500\ \text{s}},\ \text{or}\quad t = 8\ \text{min}\ 20\ \text{s}.$$

1.27 1 mL = 10^{-3} L = $10^{-3} \times 10^3$ cm^3 = 1 cm^3. $\boxed{2.5\ \text{mL} = 2.5\ \text{cm}^3}$.

1.29 Distance = speed of light $\times$ time = 3.0×10^8 m/s $\times$ 1×10^{-9} s = 0.3 m = $\boxed{30\ \text{cm}}$.

1.31 Area of one grain $\approx (0.8\ \mu\text{m})^2 = (0.8 \times 10^{-6}\ \text{m})^2 = 6.4 \times 10^{-13}\ \text{m}^2$.

$$\text{Area of one grain in cm}^2 = 6.4 \times 10^{-13}\ \text{m}^2 \times \left(\frac{100\ \text{cm}}{1\ \text{m}}\right)^2 = 6.4 \times 10^{-9}\ \text{cm}^2.$$

Number of grains in 1 cm^2 = 1 cm^2/area of 1 grain.

$$\text{Number of grains in 1 cm}^2 = \frac{1\ \text{cm}^2}{6.4 \times 10^{-9}\ \text{cm}^2} \approx \boxed{1.6 \times 10^8}.$$

1.33 Convert height to inches and then use the conversion factor of inches to cm.
Height = 5'11" = 71 in $\times$ 2.54 cm/in = $\boxed{180 \text{ cm}}$.

1.35 First compute the conversion factor between miles and km.
1 mile = 5280 ft $\times$ 12 in/ft $\times$ 2.54 cm/in $\times$ 10^{-2} m/cm = 1609 m.
1 mile = 1609 m $\times$ 10^{-3} km/m = 1.609 km.

$$150 \text{ km/h} \times \frac{1 \text{ mile}}{1.609 \text{ km}} = \boxed{93.2 \text{ miles/h}}.$$

1.37
$$\text{Cost} = \text{DM } 547 \times \frac{\$1}{\text{DM } 1.6765} = \boxed{\$\ 326.27}.$$

1.39 10 acres = 10 acre $\times$ 43,560 ft^2/acre $\times$ (12 in/ft)^2 $\times$ (2.54 cm/in)^2 $\times$ (10^{-5} km/cm)^2,
10 acres = $\boxed{4.0 \times 10^{-2} \text{ km}^2}$.

1.41 Estimate 100 bills/cm. Then

$$\text{height of 1 trillion dollars} = 10^{12} \text{ bills} \times \frac{1 \text{ cm}}{100 \text{ bills}} = 10^{10} \text{ cm} = \boxed{10^8 \text{ m}}.$$

Note: 10^8 m is approximately $\frac{1}{4}$ the distance from the earth to the moon.

1.43 We estimate the number of cars at 6×10^7 (about 1/4 the population) and the number of miles/car year is about 15000 miles. The amount of gasoline used is found from the product of the estimated number of cars times the estimated miles per car divided by the estimated average miles/gallon ≈ 15 miles/gallon.

$$\text{Amount of gasoline} = \frac{6 \times 10^7 \text{ cars} \times 1.5 \times 10^4 \text{ miles/car}}{15 \text{ miles/gallon}}$$

$\boxed{\text{Amount of gasoline} = 6 \times 10^{10} \text{ gallons.}}$

1.45 Refer to the figure.

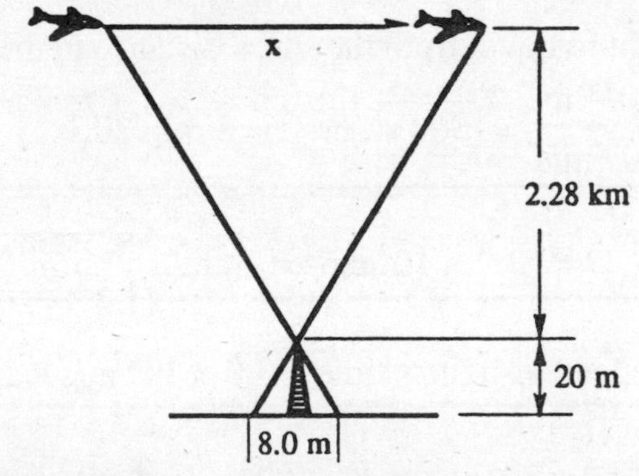

By similar triangles $\frac{8.0}{20} = \frac{x}{2.28 \text{ km}}$. $\boxed{x = 0.91 \text{ km}}$.

1.47
$$1 \text{ min} = 1 \text{ min} \times \frac{60 \text{ s}}{1 \text{ min}} \times \frac{1000 \text{ ms}}{1 \text{ s}} = \boxed{60\,000 \text{ ms}}.$$

A minute is 60 000 ms.

1.49
$$\frac{1 \text{ mg}}{1 \text{ kg}} = \frac{10^{-3} \text{ g}}{10^3 \text{ g}} = \boxed{10^{-6}}.$$

A milligram is one-millionth of a kilogram.

1.51 Watch accuracy is 15 s/mo = 0.5 s/day. It is $\boxed{\text{less accurate}}$ than the pendulum clock shown in Fig. 1.7 at $\approx 5 \times 10^{-3}$ s/day.

1.53 Rearrange the definition of an angle, $\theta = s/r$, to get $r = s/\theta$. This equation is for θ in radians. We are given θ in degrees, so we convert to radians. This gives

$$r = \frac{770\text{ km}}{7^\circ} \times \frac{360^\circ}{2\,\pi\text{ radians}} = 6300\text{ km}.$$

$$\text{Circumference} = 2\pi r = 2\pi \times 6300\text{ km} = \boxed{40\,000\text{ km}}.$$

1.55 Calculate T^2/R^3 for each planet using data of Table 1.1.

Mercury	1.0021	Mars	0.9996	Uranus	1.0000
Venus	1.0008	Jupiter	0.9990	Neptune	0.9979
Earth	1.0000	Saturn	1.0029	Pluto	1.0145

$$\%\text{ difference} = \frac{\text{highest} - \text{lowest}}{(\text{highest} + \text{lowest})/2} \times 100\%$$

$$\%\text{ difference} = \frac{1.0145 - 0.9979}{1.0145 + 0.9979} \times 2 \times 100\% = \boxed{1.6\%}.$$

1.57 Apply the conversion factors:

$$27\text{ miles/gal} \times 1.609\text{ km/mile}\left(\frac{1\text{ gal}}{3.7853\text{ L}}\right) = 11.48\text{ km/L}.$$

The inverse of this result is the answer we are looking for:

$$\frac{1}{11.48\text{ km/L}} = 0.08713\text{ L/km} = \boxed{8.7\text{ L/100 km}}.$$

1.59 The volume of the cylinder is $V = \pi r^2 h$. Compute V including the uncertainties.

$$V = \pi(r \pm \Delta r)^2(h \pm \Delta h) = \pi\left(r^2 \pm 2r\Delta r \pm (\Delta r)^2\right)(h \pm \Delta h),$$

$$V = \pi r^2 h \pm \pi r^2 \Delta h \pm 2\pi r \Delta r h \pm 2\pi r \Delta r \Delta h \pm \pi(\Delta r)^2(h \pm \Delta h).$$

When Δr and Δh are small compared to r and h, we can ignore terms containing the products $\Delta r\Delta h$ and $\Delta r\Delta r$. In that limit the volume becomes

$$\boxed{V = \pi r^2 h \pm 2\pi r h \Delta r \pm \pi r^2 \Delta h}.$$

1.61 We solve this by applying Eq. (1.2) for k corresponding to the moon and the satellite orbiting the earth. Then k is the same for both of them:

$$k = \frac{T_m^2}{R_m^3} = \frac{T_s^2}{R_s^3},$$ where m stands for moon and s for the satellite.

$$\frac{T_s^2}{T_m^2} = \frac{R_s^3}{R_m^3} = \frac{(\frac{1}{2}R_m)^3}{R_m^3} = \left(\frac{1}{2}\right)^3$$

$$\boxed{\frac{T_s}{T_m} = \left(\frac{1}{2}\right)^{3/2} = 0.354}.$$

Chapter 2

MOTION IN ONE DIMENSION

2.1 Multiply the speed by the appropriate conversion factor:

(a) $55\ \frac{\text{miles}}{\text{h}} \times 1.609\ \frac{\text{km}}{\text{mile}} = \boxed{88\ \text{km/h}}$.

(b) $55\ \frac{\text{miles}}{\text{h}} \times 1609\ \frac{\text{m}}{\text{mile}} \times \frac{1\ \text{h}}{3600\ \text{s}} = \boxed{25\ \text{m/s}}$.

2.3 Distance = speed × time = 95 km/h × 3.5 h = $\boxed{330\ \text{km}}$.

2.5 Average speed of cyclist A = $\frac{\text{distance A traveled}}{\text{time}} = \frac{30\ \text{km}}{t} = 20$ km/h.

The time spent by cyclist A was $t = \frac{30\ \text{km}}{20\ \text{km/h}} = 1.5$ h.

Average speed of cyclist B when actually riding was

Average speed of B = $\frac{\text{distance}}{\text{riding time}} = \frac{30\ \text{km}}{1.5\ \text{h} - \frac{1}{3}\ \text{h}} = \boxed{26\ \text{km/h}}$.

2.7 Linear speed in orbit can be found from the path length (the circumference of the orbit) divided by the period (one year). Thus the speed is

$$\text{linear speed} = \frac{2\pi R}{T} = \frac{2\pi \times 1.5 \times 10^{11}\ \text{m}}{3.16 \times 10^{7}\ \text{s}} = \boxed{3.0 \times 10^{4}\ \text{m/s}}.$$

Note that 1 year = $1\ \text{y} \times 365\frac{1}{4}\ \text{d/y} \times 24\ \text{h/d} \times 3600\ \text{s/h} = 3.16 \times 10^{7}$ s.

2.9 The speed of light is so large compared to the speed of sound that for distances of a few kilometers we can consider the time of travel of the light to be essentially zero. The time for sound to arrive depends on the distance and the speed of sound through $t = \frac{s}{v}$. The time for sound to travel 1 km (1000 m) is

$t = \frac{1000\ \text{m}}{340\ \text{m/s}} = 2.94$ s. $\quad t \approx \boxed{3\ \text{s}}$.

2.11 Let time of head start = t. The distance you run and the distance the dog runs must be the same.

your speed × time you run = dog's speed × time dog runs,

your speed × 30 s = dog's speed × (t + 30 s)

10 km/h × 30 s = 8.5 km/h × (t + 30 s)

300 s = 8.5 t + 255 s, $\quad$ 8.5 t = 45 s, $\quad \boxed{t = 5.3\ \text{s.}}$

2.13 The time for the train to reach the station = $t = \frac{100\ \text{km}}{70\ \text{km/h}}$.

The distance traveled by the bird = $v_{bird}\, t = 100\ \text{km/h} \times \frac{100\ \text{km}}{70\ \text{km/h}} = \boxed{140\ \text{km}}$.

2.15 The net displacement is the sum of the individual displacements. Each of these is the product of the velocity with the time. So

$s = v_1t_1 + v_2t_2 = 85\text{ km/h} \times 2.0\text{ h} - 40\text{ km/h} \times 3.0\text{ h} = \boxed{50\text{ km}}$.

2.17 The distance between the children is the distance $x = x_1 - x_2$ where x_1 is the final position of child 1 and x_2 is the final position of the child 2.

$x = x_1 - x_2 = v_1t - v_2t$

$x = [v_1 - v_2]t = [3.5\text{ m/s} - (-4.0\text{ m/s})]12\text{ s} = \boxed{90\text{ m}}$.

2.19 The total distance apart x is the sum of the two distances x_1 and x_2.

$x = x_1 + x_2 = 200$ km.

$x_1 = v_1t$, $x_2 = v_2t$, and $v_1 = 2v_2$. So $2v_2t + v_2t = 3v_2t = x$.

$v_2 = \dfrac{200\text{ km}}{3 \times 1\text{ h}} = \boxed{66.7\text{ km/h}}$. $v_1 = 2v_2 = \boxed{133\text{ km/h}}$.

2.21 From the graph: (a) $\boxed{v_o = 50\text{ km/h}}$ and (b) $\boxed{v_{40} = 110\text{ km/h}}$.

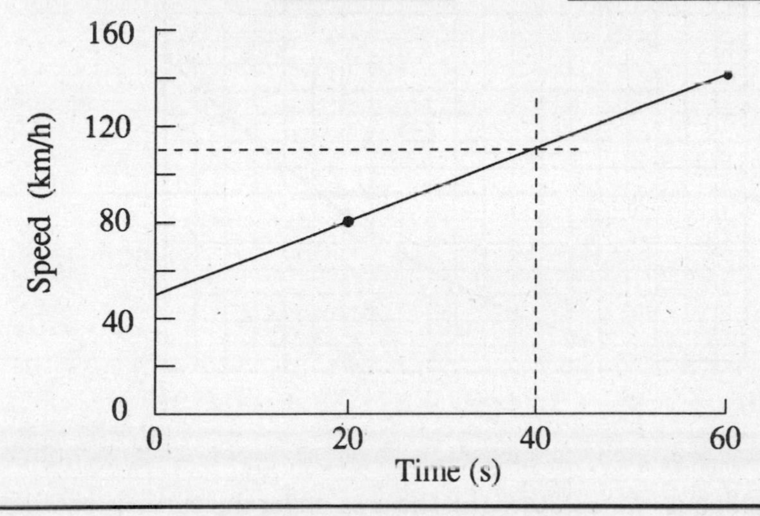

2.23 x = area under speed-time curve $= \frac{1}{2}\, 30\text{ m/s} \times 50\text{ s} = \boxed{750\text{ m}}$.

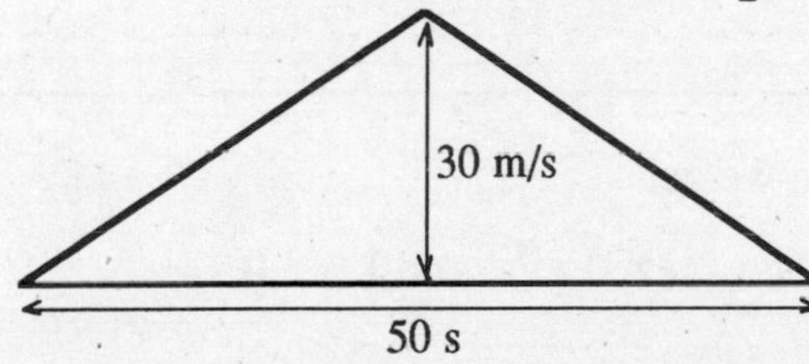

2.25 Distance = area under the line = $\frac{1}{2} \times 20$ m/s $\times$ 10 s = $\boxed{100\text{ m}}$.

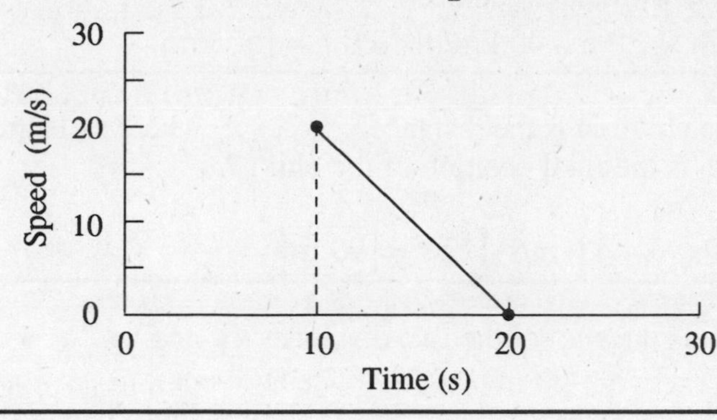

2.27 From the graph:

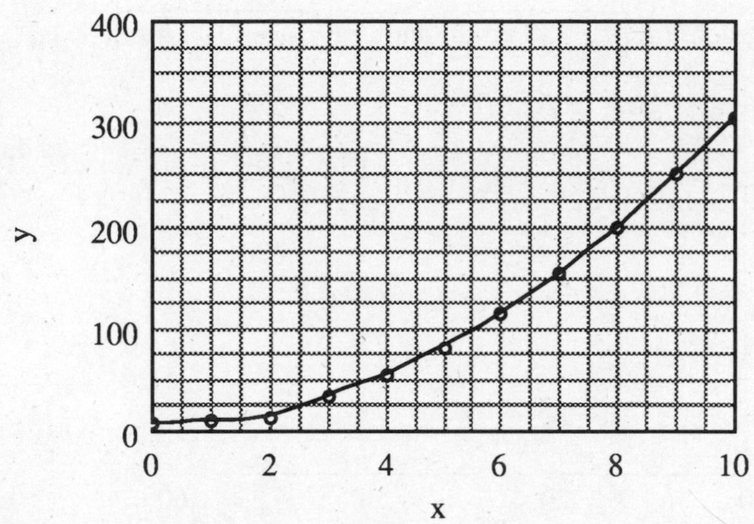

(a) At x = 5 the slope is $\frac{\Delta y}{\Delta x} = \boxed{30}$. (b) The area under the curve is obtained from counting the squares, each with an area of $25 \times \frac{1}{2} = 12.5$. The resulting area is $\boxed{1100}$.

2.29 (a) From the definition of acceleration we get

$$a = \frac{v - v_0}{t}, \text{ where } v = 17.0 \text{ m/s}, v_0 = 8.0 \text{ m/s, and } t = 3.0 \text{ s.}$$

$$a = \frac{17.0 \text{ m/s} - 8.0 \text{ m/s}}{3.0 \text{ s}} = \frac{9.0 \text{ m/s}}{3.0 \text{ s}} = \boxed{3.0 \text{ m/s}^2}.$$

(b) $x - x_0 = v_0 t + \frac{1}{2} at^2 = 8.0 \text{ m/s} \times 3.0 \text{ s} + \frac{1}{2} \times 3.0 \text{ m/s}^2 (3.0 \text{ s})^2$

$\boxed{x - x_0 = 38 \text{ m.}}$

2.31 Acceleration = $a = \frac{\Delta v}{\Delta T} = \frac{26.8 \text{ m/s}}{5.0 \text{ s}} = \boxed{5.4 \text{ m/s}^2}$.

2.33 (a) The acceleration can be obtained from the equation $v^2 - v_o^2 = 2as$.

$a = \frac{v^2 - v_o^2}{2s}$. In this case v = 0, so $a = \frac{-(v_o)^2}{2s}$. For each of the four cases we find: (1) $a = \frac{-(20\text{ mi/h})^2}{2(20\text{ ft})} = -10\ (\text{mi/h})^2/\text{ft}$. Notice that we do not need to convert to SI units in order to compare the four accelerations.

(2) $a = \frac{-(30\text{ mi/h})^2}{2(45\text{ ft})} = -10\ (\text{mi/h})^2/\text{ft}$.

(3) $a = \frac{-(40\text{ mi/h})^2}{2(80\text{ft})} = -10\ (\text{mi/h})^2/\text{ft}$.

(2) $a = \frac{-(50\text{ mi/h})^2}{2(125\text{ ft})} = -10\ (\text{mi/h})^2/\text{ft}$. $\boxed{\text{The accelerations are constant.}}$

(b) $s = \frac{-(v_o)^2}{2a} = \frac{-(60\text{ mi/h})^2}{2(-10\ (\text{mi/h})^2/\text{ft})} = \boxed{180\text{ ft}}$.

2.35 In each case you travel a distance of $v_o t$ during the reaction time t = 0.75 s.
If we choose units of ft/s for the speed the the distance will be given in feet. The units of mi/h may be converted into units of ft/s from
1 mi/h = 5280 ft/3600 s = 1.47 ft/s.
For speeds of 20 mi/h, 40 mi/h, and 50 mi/h the distances traveled during the reaction time are

$$20\text{ mi/h} \times \frac{1.47\text{ ft/s}}{\text{mi/h}} \times 0.75\text{ s} = 22\text{ ft}$$

$$40\text{ mi/h} \times \frac{1.47\text{ ft/s}}{\text{mi/h}} \times 0.75\text{ s} = 44\text{ ft}$$

$$50\text{ mi/h} \times \frac{1.47\text{ ft/s}}{\text{mi/h}} \times 0.75\text{ s} = 55\text{ ft}.$$

The total stopping distances are the distance traveled during reaction plus the distance traveled during braking (found in the table of Problem 2.29).

At 20 mi/h, stopping distance = 22 ft + 20 ft = $\boxed{42\text{ ft}}$.

At 40 mi/h, stopping distance = 44 ft + 80 ft = $\boxed{124\text{ ft}}$.

At 50 mi/h, stopping distance = 55 ft + 125 ft = $\boxed{180\text{ ft}}$.

2.37 First convert the speed to m/s.

$$48\text{ km/h} \times \frac{1000\text{ m}}{1\text{ km}} \times \frac{1\text{ h}}{3600\text{ s}} = 13.3\text{ m/s},$$

96 km/h = 26.7 m/s, and 130 km/h = 36.1 m/s.

(a) $a_1 = \frac{13.3\text{ m/s}}{3.6\text{ s}} = \boxed{3.7\text{ m/s}^2}$.

(b) $a_2 = \frac{26.7\text{ m/s} - 13.3\text{ m/s}}{10.2\text{ s} - 3.6\text{ s}} = \boxed{2.0\text{ m/s}^2}$.

(c) $\overline{a} = \frac{v^2 - v_o^2}{2s} = \frac{(36.1\text{ m/s})^2}{2 \times 400\text{ m}} = \boxed{1.6\text{ m/s}^2}$.

2.39 (a) Since $v_o = 0$, $v_{max} = a_1t_1 = 2.0\ \text{ft/s}^2 \times 12\ \text{s} = \boxed{24\ \text{ft/s}}$.

(b) With $v_o = 0$, $s_1 = \frac{1}{2}a_1t_1^2 = \frac{1}{2}(2.0\ \text{ft/s}^2)(12\ \text{s})^2 = 144\ \text{ft}$.

In the second time interval we start with an initial velocity $v_o = 24$ ft/s, so

$s_2 = v_ot_2 + \frac{1}{2}a_2t_2^2 = (24\ \text{ft/s})(12\ \text{s}) + \frac{1}{2}(-2\ \text{ft/s}^2)(12\ \text{s})^2 = 144\ \text{ft}$.

Total distance $= s_1 + s_2 = \boxed{288\ \text{ft}}$.

(c) $\overline{v} = \dfrac{\Delta s}{\Delta t} = \dfrac{288\ \text{ft}}{24\ \text{s}} = \boxed{12\ \text{ft/s}}$.

2.41 Assume the length of the side of the square field is L. Bike 1 travels the diagonal, a distance $\sqrt{2}\,L$, with constant acceleration a and initial velocity $v_o = 0$. Using the relation $s_1 = \frac{1}{2}at^2$, we get $a = \dfrac{2\,s}{t^2} = \dfrac{2\sqrt{2}\,L}{t^2}$.

Bike 2 travels a distance 2L in the same time t while traveling at constant speed v.

Using the relation $s_2 = vt$, we get $t = \dfrac{2L}{v}$.

Substitute this value of t into the equation for a to get

$$a = \frac{2\sqrt{2}\,L}{\left(\frac{2L}{v}\right)^2} \text{ or } \boxed{a = \frac{v^2}{\sqrt{2}L}}.$$

2.43 The relationship we need is $x = v_ot + \frac{1}{2}at^2$, where $v_o = 0$. Then the time to fall is given by $t = \sqrt{\dfrac{2x}{a}} = \sqrt{\dfrac{2 \times 54.6\ \text{m}}{9.80\ \text{m/s}^2}} = \boxed{3.34\ \text{s}}$.

2.45 Velocity $= v = v_o + at$. Here $v_o = 0$, so $v = at = 9.8\ \text{m/s}^2 \times t$.

Distance $= x - x_o = v_ot\ \frac{1}{2}at^2 = \frac{1}{2}9.8\ \text{m/s}^2 \times t^2$, since $v_o = 0$.

t (s)	v (m/s)	$x - x_o$ (m)
0.0	0.0	0.0
0.5	4.9	1.2
1.0	9.8	4.9
1.5	14.7	11.0
2.0	19.6	19.6

2.47 Call the starting point 0. Then the distance is given by $x = v_ot + \frac{1}{2}at^2$. Since the ball starts from rest, $v_o = 0$. Then $x = \frac{1}{2}at^2$.

$x_2(t = 4.0\ \text{s}) = 1.60\ \text{m} = \frac{1}{2}a\,(4.0\ \text{s})^2$,

$a = \dfrac{2\,x}{t^2} = \dfrac{2 \times 1.60\ \text{m}}{(4.0\ \text{s})^2} = 0.20\ \text{m/s}^2$.

$x_1(t = 2.0\ s) = \frac{1}{2} \times 0.20\ m/s^2 \times (2.0\ s)^2 = \boxed{0.40\ m}$.

$x_4(t = 8.0\ s) = \frac{1}{2} \times 0.20\ m/s^2 \times (8.0\ s)^2 = \boxed{6.4\ m}$.

2.49 The average velocity over an interval is $\overline{v} = \frac{\Delta x}{\Delta t}$. Each interval Δt = 1/rate = 0.25 s.

$$\overline{v_1} = \frac{0.70\ m}{0.25\ s} = 2.80\ m/s,$$

$$\overline{v_2} = \frac{0.90\ m}{0.25\ s} = 3.60\ m/s,$$

$$\overline{v_3} = \frac{1.10\ m}{0.25\ s} = 4.40\ m/s,$$

$$\overline{v_4} = \frac{1.30\ m}{0.25\ s} = 5.20\ m/s.$$

The average acceleration is $\overline{a} = \frac{\Delta v}{\Delta t}$.

$$\overline{a_{12}} = \frac{3.60\ m/s - 2.80\ m/s}{0.25\ s} = 3.2\ m/s^2,$$

$$\overline{a_{23}} = \frac{4.40\ m/s - 3.60\ m/s}{0.25\ s} = 3.2\ m/s^2,$$

$$\overline{a_{34}} = \frac{5.20\ m/s - 4.40\ m/s}{0.25\ s} = 3.2\ m/s^2. \qquad \boxed{a = 3.2\ m/s^2.}$$

2.51 Time to fall 0.50 m from rest is $t = \sqrt{\frac{2x}{g}} = \sqrt{\frac{2 \times 0.5\ m}{9.8\ m/s^2}} = 0.32\ s.$

The cylinder must rotate $\frac{1}{2}$ turn in 0.32 s, so it takes T = 0.64 s to make a complete revolution. The number of revolutions per second is $\frac{1}{T}$.

The number of revolutions per second $= \frac{1}{0.64\ s} = \boxed{1.6\ rev/s}$.

2.53 The total distance traveled is given by the average speed times the time: $s = \overline{v}\, t = 75\ km/h \times 3.0\ h = 225\ km$. The distance traveled during the first hour is $s_1 = 90\ km/h \times 1.0\ h = 90\ km$. The distance traveled during the remainder of the trip is $s_2 = 225\ km - 90\ km = 135\ km$.

Average speed of the remainder of the trip $= \frac{135\ km}{2.0\ h} = \boxed{68\ km/h}$.

2.55 (a) The average speed is the distance divided by the time;

$$\overline{v_c} = \frac{s}{t} = \frac{5564\ km}{3\ h} \approx \boxed{1900\ km/h}.$$

(b) $\overline{v_w} = \frac{120\ ft}{12\ s} \times \frac{0.305\ m}{ft} \times \frac{10^{-3}\ km}{m} \times \frac{3600\ s}{h} = \boxed{11\ km/h}$.

(c) The ratio of the speeds is $\frac{v_c}{v_w} = \frac{1900}{11} = \boxed{170}$.

2.57 (a) Convert the units: $60\ \text{mi/h} \times 1609\ \text{m/mi} \times \frac{1\ \text{h}}{3600\ \text{s}} = 26.8\ \text{m/s} \approx \boxed{27\ \text{m/s}}$.

(b) Average acceleration is $\overline{a} = \frac{\Delta v}{\Delta t} = \frac{26.8\ \text{m/s}}{6.5\ \text{s}} = \boxed{4.1\ \text{m/s}^2}$.

(c) Distance traveled is $x - x_o = \frac{1}{2}at^2 = \frac{1}{2}(4.1\ \text{m/s}^2)(6.5\ \text{s})^2 = \boxed{87\ \text{m}}$.

2.59 Each bug travels the same distance R in the same time t.

For bug (1), $R = v_1t$. For bug (2), $v_o = 0$, so $R = \frac{1}{2}at^2$.

For bug (3), $R = v_2t - \frac{1}{2}at^2$.

(a) Rearrange the equation for bug (3) to get

$v_2t = R + \frac{1}{2}at^2 = R + R = 2R = 2v_1t$,

$\boxed{v_2 = 2\ v_1.}$

(b) Combining the equations for bug (1) and bug (2) we get

$v_1t = \frac{1}{2}at^2$, so $a = \frac{2v_1t}{t^2} = \frac{2v_1}{t}$

But $t = \frac{R}{v_1}$, which leads to $a = \frac{2v_1}{R/v_1} = \boxed{\frac{2v_1^2}{R}}$.

2.61 If dropped from height x_o above the ground, the ball would have an initial velocity $v_o = 0$. We call the ground position $x = 0$.

The equation for the position becomes $x - x_o = \frac{1}{2}at_1^2$. The acceleration is $a = -g = -9.8\ \text{m/s}^2$. Inserting the numbers we get

$-20\ \text{m} = \frac{1}{2}(-g)t_1^2$. Solving for t_1, the time to fall, we get $t_1 = 2.02\ \text{s}$.

If thrown upward with initial speed v_o so that ball just reaches the top with speed $v = 0$, then the upward speed is found from

$v^2 - v_o^2 = 0^2 - v_o^2 = 2a(x - x_o) = 2(-9.8\ \text{m/s}^2)(20\ \text{m} - 0)$,

$v_o^2 = 392\ \text{m}^2/\text{s}^2$,

$v_o = 19.8\ \text{m/s}^2$. Note that we do not round off the number at this point because we want to avoid round off errors in the subsequent computations.

If the ball were thrown downward from building with speed v_o, then the equation for the displacement is

$x - x_o = -20\ \text{m} = v_ot_2 + \frac{1}{2}at_2^2 = (-19.8\ \text{m/s})t_2 + \frac{1}{2}(-9.8\ \text{m/s}^2)t_2^2$. Upon solving for t_2 we get $t_2 = 0.84$ s. Thus the ball reaches the ground sooner when thrown by an amount $\Delta t = t_1 - t_2 = 1.18\ \text{s} \approx \boxed{1.2\ \text{s}}$.

2.63 First we calculate the velocity of the ball bearing as it reaches the 10 m mark.

$v^2 = v_o^2 + 2ax$, with $v_o = 0$, $x = -10$ m, and $a = -9.8\ \text{m/s}^2$.

$v = \sqrt{2 \times 9.8 \times 10} = \pm 14\ \text{m/s}$. (We choose the negative square root since v is downward.) Call time $t = 0$ the time of release of the second bearing and the time that the first bearing has fallen 10 m. Measure distances from the position of the first bearing at $t = 0$.

(a) Find the time for the two ball bearings to reach the same position.

$x_1 = x_{1o} + v_{1o}t - \frac{1}{2}gt^2 = x_2 = x_{2o} + v_{2o}t - \frac{1}{2}gt^2,$

$0 + (-14\text{ m/s})t - \frac{1}{2}gt^2 = -1\text{ m} + 0 - \frac{1}{2}gt^2 \quad \text{or} \quad t = \frac{-1\text{m}}{-14\text{ m/s}} = \boxed{0.071\text{ s}}.$

(b) $x_2 - x_{2o} = v_{2o}t - \frac{1}{2}gt^2 = 0 - \frac{1}{2}9.8\text{ m/s}^2\,(0.071\text{ s})^2 = \boxed{0.025\text{ m}}.$

2.65 We can relate the final velocity to the height x_o through $v_f^2 = v_o^2 + 2a(x - x_o)$. If $v_o = 0$, $x = 0$, and $a = -g$, then we get $x_o = \frac{v_f^2}{2g}$. We also know that $x - x_o = -\frac{1}{2}gt^2$. By combining these equations (with $x = 0$) we find that $t = \frac{v_f}{g}$. The distance traveled by the car in time t is $x_{car} = v_{car}t = v_ft = v_f^2/g = 2x_o$. x_o is the height of the tower: $x_o = x_{tower}$. Thus $\boxed{x_{car} = 2\,x_{tower}}$.

2.67 This problem is described with Eq. (2.7): $x = x_o + v_ot + \frac{1}{2}at^2$. We rearrange this equation to get: $\frac{1}{2}at^2 + v_ot + (x_o - x) = 0$. Upon comparing this with the quadratic formula we find that the solutions are:

$$t = \frac{-v_o \pm \sqrt{v_o^2 - 4(\frac{1}{2}a)(x_o - x)}}{2\,(\frac{1}{2}a)}.$$

The distance $x - x_o = 1700$ m, so $x_o - x = -1700$ m. When this and the appropriate values are inserted for $v_o = 4.5$ m/s,and $a = 1.8\text{ m/s}^2$, we find

$$t = \frac{-4.5 \pm \sqrt{(4.5)^2 - 4(\frac{1}{2}1.8)(-1700)}}{2\,(\frac{1}{2}1.8)}\text{ s} = \frac{-4.5 \pm 78.4}{1.8}\text{ s.}$$

Thus we have two possible answers, $t = 41$ s, or $t = -46$ s. We choose the value corresponding to a positive elapsed time, thus the time is $\boxed{t = 41\text{ s}}$.

Chapter 3

MOTION IN TWO DIMENSIONS

3.1 (a) First make a drawing.

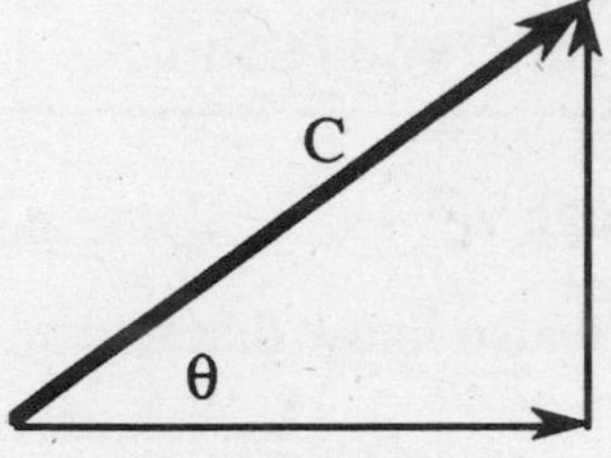

Using the Pythagorean theorem, we find the length of the vector from the starting point to the ending point: $C^2 = A^2 + B^2 = 6^2 + 8^2 = 100$ blocks^2.

$\boxed{C = 10 \text{ blocks.}}$ (b) $\tan\theta = \frac{A}{B} = \frac{6}{8}$, $\boxed{\theta = 37° \text{ north of east.}}$

3.3 (a) By graphical addition we find $\boxed{\mathbf{A} + \mathbf{B} = 32 \text{ units at } 306°}$.

(b) Similarly we find $\boxed{\mathbf{A} - \mathbf{B} = 28 \text{ units at } 177°}$.

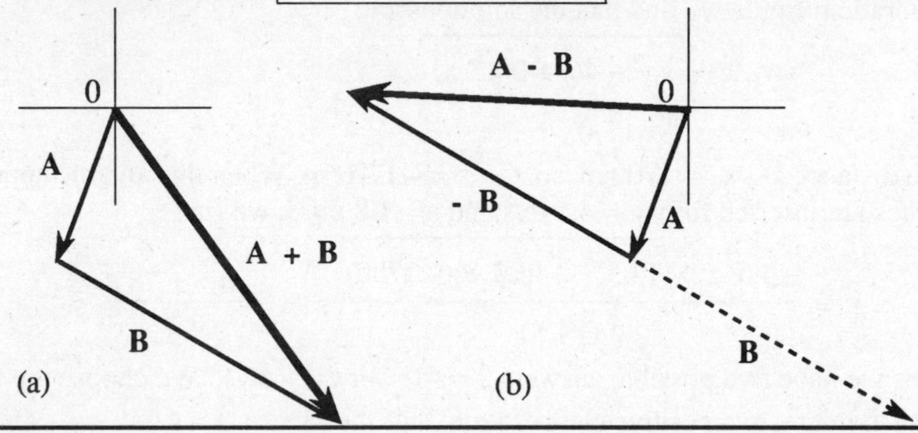

3.5

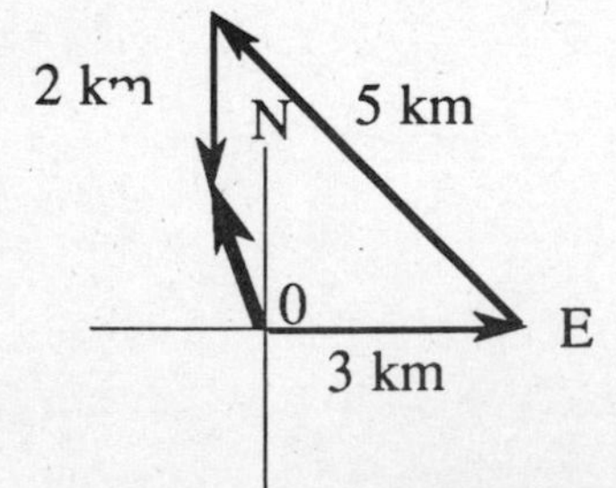

Using graphical analysis we find:

a distance of $\boxed{1.6 \text{ km}}$ and a direction of $\boxed{109° \text{ from E}}$.

3.7 (a) $\boxed{A \text{ at } 0°}$. (b) $\boxed{0}$. (c) $\boxed{1.73\, A \text{ at } 90°}$. (d) $\boxed{2\mathbf{A}}$.

3.9 The two segments of the trip are at right angles and form the sides of a right triangle. The direct distance from start to finish is the hypotenuse of the triangle. Use the Pythagorean theorem $C^2 = A^2 + B^2$, where C = 112 m and B = 2A.
$112^2 = A^2 + (2A)^2 = 5A^2$.
$A = \frac{112\text{ m}}{\sqrt{5}} = \boxed{50\text{ m}}$. $B = 2A = \boxed{100\text{ m}}$.
The angle from north is obtained from $\theta = \tan^{-1}\frac{100}{50} = \boxed{63°}$.

3.11 The x component is $A_x = A\cos\theta = 12\cos 50° = \boxed{7.7}$ and the y component is $A_y = A\sin\theta = 12\sin 50° = \boxed{9.2}$.

3.13 The magnitude is $A = \sqrt{A_x^2 + A_y^2} = \sqrt{4.5^2 + (-6.8)^2} = \boxed{8.2}$.
The angle with the x axis is $\theta = \tan^{-1}\frac{A_y}{A_x} = \tan^{-1}\frac{-6.8}{4.5} = \boxed{-57°}$.

3.15 The y component of the vector is $A_y = A\sin\theta$, where $\theta = 127°$. The magnitude is
$A = \frac{A_y}{\sin\theta} = \frac{18.0}{\sin 127°} = \frac{18.0}{0.799} = \boxed{22.5}$. The x–component is
$A_x = A\cos\theta = \frac{A_y\cos\theta}{\sin\theta} = \frac{A_y}{\tan\theta} = \frac{18.0}{\tan 127°} = \boxed{-13.6}$.

3.17 First resolve the vectors into x and y components and then add to get
$C_x = A_x + B_x = 8.0\cos 60° + 6.0\cos(-30°) = 4 + 5 = 9$.
$C_y = A_y + B_y = 8.0\sin 60° + 6.0\sin(-30°) = 7 + (-3) = 4$.
$C = \sqrt{C_x^2 + C_y^2} = \sqrt{9^2 + 4^2} = \boxed{10}$. $\theta = \tan^{-1}\left(\frac{4}{9}\right) = \boxed{24°}$.

3.19 First resolve the vectors into x and y components. Then subtract the components of **B** from twice the components of **A** to get
$D_x = 2A_x - B_x = 2 \times 8.0\cos 60° - 6.0\cos(-30°) = 8 - 5 = 3$.
$D_y = 2A_y - B_y = 2 \times 8.0\sin 60° - 6.0\sin(-30°) = 14 - (-3) = 17$.
$D = \sqrt{D_x^2 + D_y^2} = \sqrt{3^2 + 17^2} = \boxed{17}$. $\theta = \tan^{-1}\left(\frac{17}{3}\right) = \boxed{80°}$.

3.21 (a) Net velocity upstream is the still-water speed of the boat less the speed of the river. So v = 10.0 km/h – 3.0 km/h = 7.0 km/h. The time to go 7 km is
$t = \frac{7.0\text{ km}}{7.0\text{ km/h}} = \boxed{1.0\text{ h}}$.
(b) Going downstream the speeds add so that the time becomes
$t = \frac{7.0\text{ km}}{13.0\text{ km/h}} = \boxed{0.54\text{ h}}$.

3.23 Draw a diagram of the situation. $\mathbf{v}_w$ represents the wind velocity and $\mathbf{v}_p$ represents the velocity of the plane in the absence of the wind.

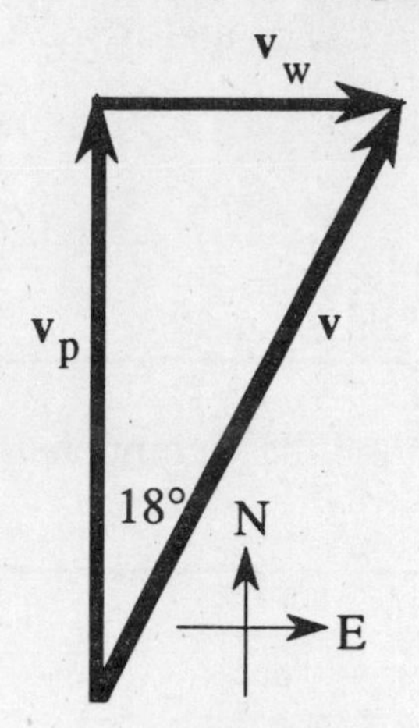

v is the velocity of the plane in the air.
From the diagram we see that
$v_w = v \sin\theta = 200 \text{ km/h} \times \sin 18° = \boxed{61.8 \text{ km/h.}}$

3.25 First make a diagram of the vector velocities.

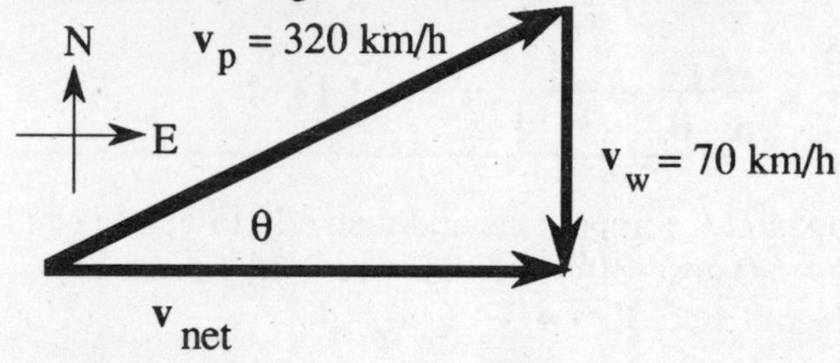

V_p is the maximum airspeed of the plane. When the cross wind starts, the pilot must turn to head somewhat into the wind in order to stay on the east bound course. We can compute the angle θ from

$\sin\theta = \frac{v_w}{v_p} = \frac{70}{320} = 0.219.$ $\theta = 12.6°.$ The net speed to the east is

$v_{net} = v_p \cos\theta = 320 \text{ km/h} \cos 12.6° = 312 \text{ km/h}.$ The time required to travel 1590 km at a speed of 312 km/h is
$t = \text{distance/speed} = 1590 \text{ km}/(312 \text{ km/h}) = 5.09 \text{ h} \approx 5.1 \text{ h}.$
The answer is $\boxed{\text{no}}$. The plane will not be on schedule.

3.27 Make a diagram of the vectors.

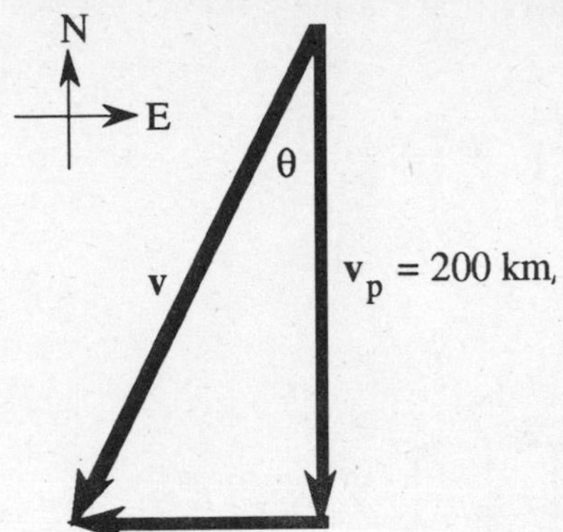

The net speed of the plane is $v = \sqrt{200^2 + 10^2}$ km/h ≈ 200 km/h.
In 2 h the plane travels 2 × 200 km/h = $\boxed{400 \text{ km}}$.
The plane travels in a direction that is an angle θ to the west of south, where θ is given by $\theta = \tan^{-1}\frac{10}{200} = \boxed{2.9°}$.

3.29 (a) Choose north as the positive direction. Then, the relative velocity seen from car B is $\mathbf{v}_{A} - \mathbf{v}_{B}$ = 90 km/h –(–70 km/h) = $\boxed{160 \text{ km/h}}$.
(b) The relative velocity from car A is
$\mathbf{v}_{B} - \mathbf{v}_{A}$ = –70 km/h –(90 km/h) = $\boxed{-160 \text{ km/h}}$.
(c) v_{AC} = 90 km/h – 100 km/h = $\boxed{-10 \text{ km/h}}$.
v_{BC} = – 70 km/h – 100 km/h = $\boxed{-170 \text{ km/h}}$.

3.31 (a) The vertical motion can be computed independently of any horizontal motion.
$y - y_o = v_{yo}t - \frac{1}{2}gt^2$. Here $v_{yo} = 0$. So,

$$t = \sqrt{\frac{2(y - y_o)}{-g}} = \sqrt{\frac{2 \times -1.20 \text{ m}}{-9.81 \text{ m/s}^2}} = \boxed{0.495 \text{ s}}.$$

(b) The coin has no horizontal motion relative to the train so it lands directly beneath the point where it was dropped.
(c) x = vt = (250 km/h)(1000m/km)(1 h/3600 s)(0.495 s) = $\boxed{34.4 \text{ m}}$.

3.33 The time is from the horizontal distance divided by the horizontal speed.

$$t = \frac{x}{v_{xo}} = \frac{17.0 \text{ m}}{9.5 \text{ m/s}} = 1.79 \text{ s}.$$

The height y_o is obtained from $y - y_o = v_{yo}t - \frac{1}{2}gt^2$. Since $v_{yo} = 0$, if we call the ground level y = 0, then the height y_o is given by
$y_o = \frac{1}{2} 9.8 \text{ m/s}^2 \times (1.79 \text{ s})^2 = \boxed{15.7 \text{ m}}$.

3.35 Make a table of t, x, and y.

t	$x = (2 + 3t + 2t^2)$ cm	$y = (2t + 5t^2)$ cm
0	2	0
1	7	7
2	16	24
3	29	51
4	46	88

Graph this data.

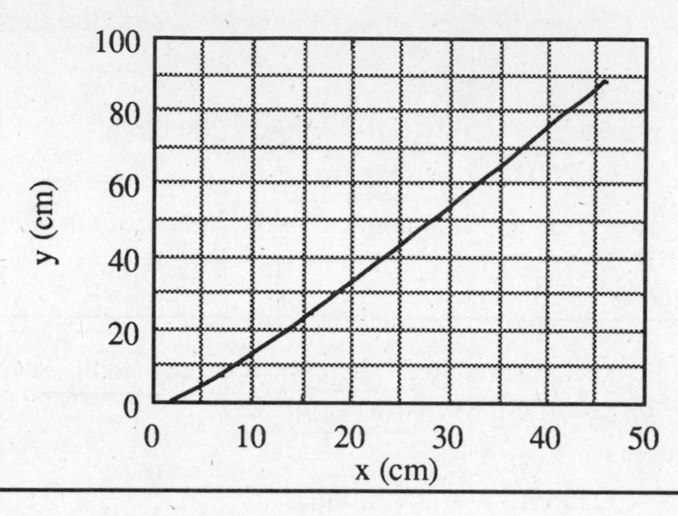

3.37 Since the arrow struck at an angle of 45° at the end of three seconds, then $v_y = v_x = v_o$. (v_x is constant at v_o.)

(a) $v_y = v_{yo} - gt = 0 - 9.80\ \text{m/s}^2 \times 3.00\ \text{s} = -29.4$ m/s.

Thus the speed with which the arrow was launched is $\boxed{29.4\ \text{m/s}}$.

(b) If we choose the ground level as y = 0, then the height of launch is y_o,

$y - y_o = v_{yo} - \frac{1}{2}gt^2$.

But v_{yo} is also 0. The equation for y_o becomes

$y_o = \frac{1}{2}gt^2 = 4.90\ \text{m/s}^2 \times (3.00\ \text{s})^2 = \boxed{44.1\ \text{m}}$.

3.39 (a) Find the time to fall from the equation $y = y_o + v_{yo}t - \frac{1}{2}(9.81)t^2$, where y = 0, $y_o = 1.20$ m, and $v_{yo} = 0$.

$$t = \sqrt{\frac{2 \times 1.20}{9.81}} = 0.495\ \text{s}.$$

The horizontal distance is $v_x t = 38.0\ \text{m/s} \times 0.495\ \text{s} = \boxed{18.8\ \text{m}}$.

(b) This is a simple range problem. $R = \dfrac{v_o^2 \sin 2\theta}{g}$, so

$$\sin 2\theta = \frac{gR}{v_o^2} = \frac{9.81\ \text{m/s}^2 \times 39.0\ \text{m}}{(38.0\ \text{m/s})^2} = 0.265.$$

$2\theta = \sin^{-1}0.265 = 15.36°$. $\boxed{\theta = 7.68°.}$

3.41 (a) There is no acceleration in the x direction so we can write
$x = v_x t = (10.0 \text{ cm/s})(5.0 \text{ s}) = 50 \text{ cm}$. In the y direction we must include the acceleration a_y. Then the y position is
$y = v_{yo}t + \frac{1}{2} a_y t^2 = (-5.0 \text{ cm/s})(5.0\text{s}) + \frac{1}{2}(4.0 \text{ cm/s}^2)(5.0 \text{ s})^2 = 25 \text{ cm}.$
The distance from the origin is just the magnitude of the displacement:
$r = \sqrt{x^2 + y^2} = \sqrt{50^2 + 25^2} \text{ cm} = \boxed{56 \text{ cm}}$.

(b) The direction can be found from $\theta = \tan^{-1}\frac{y}{x} = \tan^{-1}\frac{1}{2} = \boxed{26.6°}$.

(c) At t = 5.0 s, the instantaneous velocity has components $v_x = 10.0$ cm/s and $v_y = v_{yo} + at = -5 \text{ cm/s} + (4 \text{ cm/s}^2)(5.0 \text{ s}) = 15 \text{ cm/s}$.
The direction of the instantaneous velocity is

$$\phi = \tan^{-1}\frac{v_y}{v_x} = \tan^{-1}\frac{15 \text{ cm/s}}{10 \text{ cm/s}} = \tan^{-1}1.5 = \boxed{56°}.$$

3.43 Since range $= R = \frac{v_o^2}{g} \sin 2\theta$, the velocity can be found from

$$v_o = \sqrt{\frac{Rg}{\sin 2\theta}} = \boxed{47.1 \text{ m/s}}.$$

3.45 Since range $= R = \frac{v_o^2}{g} \sin 2\theta$, the velocity can be found from

$$v_o = \sqrt{\frac{Rg}{\sin 2\theta}} = \boxed{2.9 \text{ m/s}}.$$

3.47 Use the range formula to get

$$\Delta R = R_{45} - R_{30} = \frac{v_o^2}{g}(\sin 90° - \sin 60°) = \frac{(75 \text{ m/s})^2}{9.8 \text{ m/s}^2}(1 - .866)$$

$\boxed{\Delta R = 77 \text{ m.}}$

3.49 First we need to find the initial velocity v_o. Use $v^2 - v_o^2 = 2(-g)h$ where h = 20 m when v = 0. Thus $v_o^2 = 2gh$.
The maximum range is $R_{max} = \frac{v_o^2}{g} = 2h = \boxed{40 \text{ m}}$.

3.51 The simple range formula does not apply here. First we must find the initial vertical velocity and use it to calculate the time for the ball to reach the ground. Then we can find the horizontal distance.
$v_{oy} = 14.3 \text{ m/s} \sin 27° = 6.49 \text{ m/s}$. $v_{ox} = 14.3 \text{ m/s} \cos 27° = 12.7 \text{ m/s}$.
$y - y_o = -3.50 \text{ m} = v_{oy}t - \frac{1}{2}gt^2 = (6.49 \text{ m/s})t - (4.90 \text{ m/s}^2)t^2$.
Use the quadratic formula to get t = 1.73 s. Now compute the horizontal distance from $x = v_{xo}\, t = 12.7 \text{ m/s} \times 1.73 \text{ s} = \boxed{22.0 \text{ m}}$.

3.53 At maximum range, $\theta = 45°$ and $v_{yo} = v_{xo} = \frac{v_o}{\sqrt{2}}$. The time to travel the whole distance is $T = \frac{R_{max}}{v_{xo}} = \frac{\sqrt{2}R_{max}}{v_o}$, and $R_{max} = \frac{v_o^2}{g} = \frac{2v_{xo}^2}{g}$. The time required to go $\frac{3}{4}R_{max}$ is $t = \frac{3}{4}T$. The height at $t = \frac{3}{4}T$ is

$$y = v_{yo}t - \frac{1}{2}gt^2 = v_{yo}\frac{3T}{4} - \frac{1}{2}g\left(\frac{3T}{4}\right)^2 = v_{yo}\frac{\frac{3}{4}R_{max}}{v_{xo}} - \frac{1}{2}g\left(\frac{\frac{3}{4}R_{max}}{v_{xo}}\right)^2$$

$$y = \frac{3}{4}R_{max} - \frac{1}{2}\left(\frac{2v_{xo}^2}{R_{max}}\right)\left(\frac{9R_{max}^2}{16v_{xo}^2}\right) = R_{max}\left(\frac{3}{4} - \frac{9}{16}\right) \qquad \boxed{y = \frac{3}{16}R_{max}}.$$

3.55 The guy wire of length L is the hypotenuse of a right triangle with angle = 53° and opposite side of length h = 5.50 m. Using the definition of the sine we get

$\sin\theta = \frac{h}{L}$, so $L = \frac{h}{\sin\theta}$. $L = \frac{5.50 \text{ m}}{\sin 53°} = \boxed{6.89 \text{ m}}$.

3.57 Use a right triangle with one side h = distance from line to tower and the perpendicular side x = one half of the base line distance = 50.0 m. Then

$\tan 85.5° = \frac{h}{x}$ or $h = x \tan 85.5° = 50.0 \text{ m} \times 12.7 = \boxed{635 \text{ m}}$.

3.59 The vectors shown if (a) below can be translated to the positions shown in (b) with a net sum of zero.

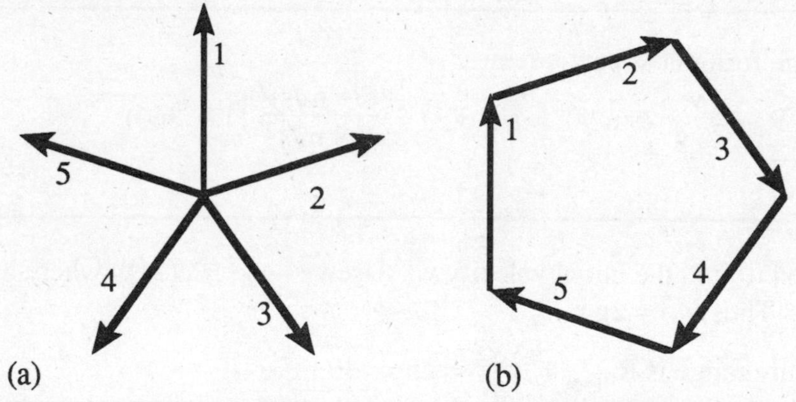

3.61 $R_{max} = \frac{v_o^2}{g}$ and $R = \frac{v_o^2 \sin 2\theta}{g}$. R_{max} occurs for $\theta = 45°$. For $\theta = 35°$, we have $R = (\sin 70°)R_{max}$ or

$$\frac{R_{max} - R}{R_{max}} = 1 - \sin 70° = 1 - 0.940 = 0.06 \text{ or } \boxed{6\%}.$$

3.63 Construct a table of the angle in degrees and in radians and of the sine of the angle.

$\theta(°)$	θ(rad)	$\sin\theta$	$\dfrac{\theta - \sin\theta}{\sin\theta}$
1	0.0174433	0.0174524	5.07×10^{-5}
3	0.0523599	0.0523360	4.57×10^{-4}
6	0.1047198	0.1045285	1.83×10^{-3}
9	0.1570796	0.1564345	4.12×10^{-3}
12	0.2094395	0.2079117	7.35×10^{-3}
15	0.2617994	0.2588190	1.15×10^{-2}
18	0.3141593	0.3090170	1.66×10^{-2}
21	0.3665191	0.3583679	2.27×10^{-2}

$\sin\theta \approx \theta$ to within 0.5 % for θ less than about 10°.

3.65 The time T to travel the whole distance is the distance divided by the horizontal speed. So $T = \dfrac{R}{v_o\cos\theta}$, but R can be found from the range formula;

$R = \dfrac{v_o^2\sin 2\theta}{g} = \dfrac{2v_o^2\sin\theta\cos\theta}{g}$, where we have used $\sin 2\theta = 2\sin\theta\cos\theta$.

The maximum height is found from $y_{max} = h = v_o(\sin 0)t - \frac{1}{2}gt^2$, where $t = \frac{1}{2}T$.

$$h = v_o\sin\theta\,\frac{R}{2v_o\cos\theta} - \frac{1}{2}\,\frac{2v_o^2\sin\theta\cos\theta}{R}\left(\frac{R}{2v_o\cos\theta}\right)^2$$

$$h = \frac{1}{2}R\tan\theta - \frac{1}{4}R\tan\theta = \boxed{\frac{1}{4}R\tan\theta}.$$

Chapter 4

FORCE AND MOTION

4.1 (a) $F = ma = (0.50\text{ kg})(4.0\text{ m/s}^2) = \boxed{2.0\text{ N}}$.

(b) $a = \dfrac{F}{m} = \dfrac{2.0\text{ N}}{20\text{ kg}} = \boxed{0.10\text{ m/s}^2}$.

4.3 $a = \dfrac{\Delta v}{\Delta t} = \dfrac{30.0 \times 10^3\text{ m/h} \times (1\text{ h}/3600\text{ s})}{5.00\text{ min} \times 60\text{ s/min}} = 2.78 \times 10^{-2}\text{ m/s}^2$.

$F = ma = 1.20 \times 10^6\text{ kg} \times 2.78 \times 10^{-2}\text{ m/s}^2 = \boxed{3.33 \times 10^4\text{ N}}$.

4.5 The acceleration is $a = \dfrac{F}{m}$, and the distance traveled is $x = \frac{1}{2}at^2$, where $v_{xo} = 0$.

$x = \frac{1}{2}\left(\dfrac{980\text{ N}}{1500\text{ kg}}\right)(5.0\text{ s})^2 = \boxed{8.2\text{ m}}$.

4.7 For a 1.0-kg object, a force of x newtons gives an acceleration of x m/s^2. So we draw a graph of acceleration versus time.

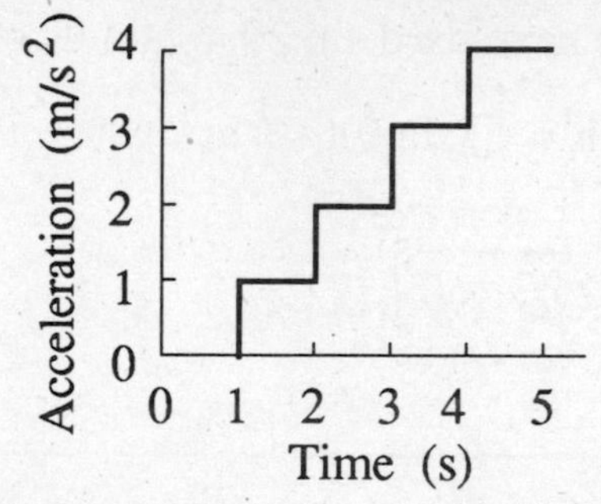

From this we get a graph of the velocity and then a graph of displacement

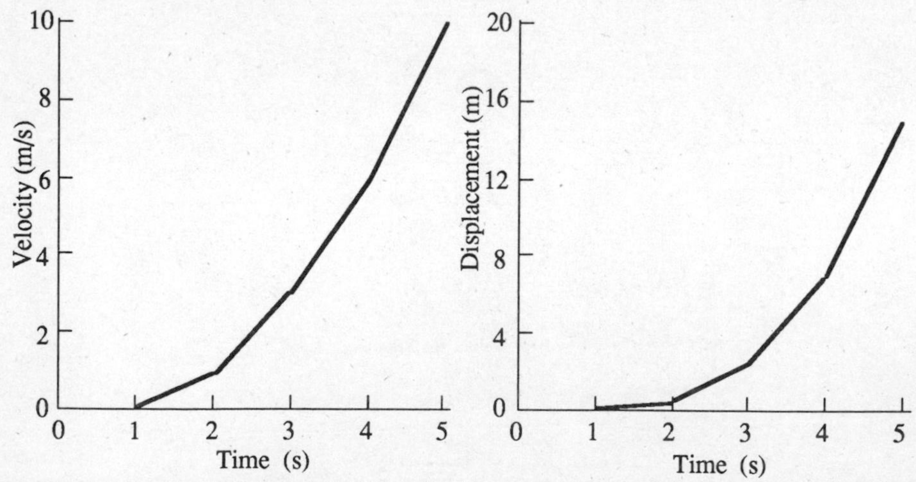

4.9 (a) 1 N = 0.2248 lb, so $1 \text{ lb} = \frac{1}{0.2248} \text{ N} = 4.448 \text{ N}$.

$150 \text{ lb} \times 4.448 \text{ N/lb} = \boxed{667 \text{ N}}$.

(b) $m = \frac{\text{weight}}{g} = \frac{667 \text{ N}}{9.81 \text{ m/s}^2} = \boxed{68.0 \text{ kg}}$.

4.11 $\text{Weight} = mg = 9.00 \text{ kg} \times 9.81 \text{ m/s}^2 = \boxed{88.2 \text{ N}}$.

4.13 $\text{Percent change} = \frac{\text{change}}{\text{initial value}} 100\% = \frac{0.01}{9.81} 100\% = \boxed{0.1\%}$.

4.15 The initial acceleration is the net force divided by the mass. The net force is the difference between the upward thrust and the downward force of gravity.

$$a = \frac{F - mg}{m} = \frac{F}{m} - g = \frac{7.4 \times 10^6 \text{ N}}{5.4 \times 10^5 \text{ kg}} - 9.8 \text{ m/s}^2 = (13.7 - 9.8) \text{ m/s}^2.$$

$\boxed{a = 3.9 \text{ m/s}^2.}$

4.17 (a) The reaction force is the force of the table on the book and is equal to the weight of the book. (b) The reaction force is the force of the book on the earth and is equal to the weight of the book. (c) The reaction force is force of the table on the book and is less than the weight of the book. The reaction force to the gravity force is the force of the book on the earth and is equal to the weight of the book.

4.19 The traction force is 32 N, so the downward force on the pulley is 32 N. The basket is supported by the force at A and the force of the rope attached to the ceiling. These forces are the same, so $F = (32 \text{ N})/2 = \boxed{16 \text{ N}}$.

4.21 (a) Scale reading momentarily increases, then returns to initial reading. (b) Scale reading same as initial reading. (c) Same as (a).

4.23 A new graph may be computed of velocity versus time.

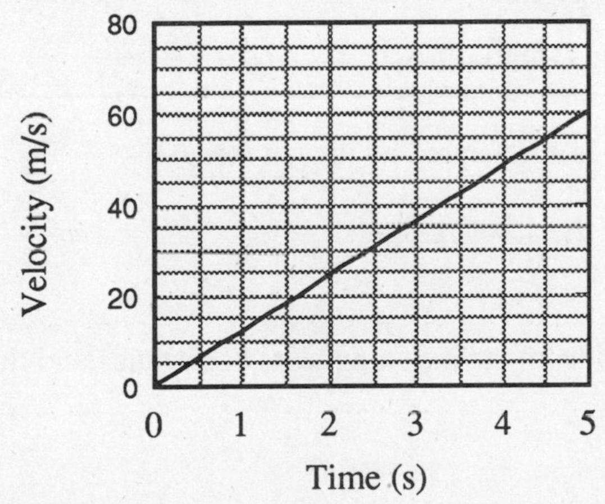

The velocity graph is linear, so the acceleration is constant. From the slope of the graph the acceleration is $a = \dfrac{60\text{ m/s} - 0}{5\text{ s}} = 12\text{ m/s}^2$.

The force is constant at $F = ma = \boxed{1.2 \times 10^4\text{ N}}$.

4.25 (a) From Newton's second law, the acceleration of the combination of masses is

$$a = \frac{F}{m_1 + m_2} = \frac{5.0\text{ N}}{2.0\text{ kg} + 0.50\text{ kg}} = \boxed{2.0\text{ m/s}^2}.$$

(b) The tension in the string provides the force to accelerate the mass m_1, so

$T = m_1 a = 2.0\text{ kg} \times 2.0\text{ m/s}^2 = \boxed{4.0\text{ N}}$.

4.27 First find the acceleration from $v^2 - v_o^2 = 2ax$. Here $v = 0$, $v_o = 200$ m/s, and $x =$ 0.030 m. So $a = \dfrac{0 - (200\text{ m/s})^2}{2 \times 0.030\text{ m}} = \dfrac{-40000\text{ m}}{0.060\text{ s}^2} = -6.67 \times 10^5\text{ m/s}^2$.

The force is $F = ma = 0.0050\text{ kg} \times 6.67 \times 10^5\text{ m/s}^2 = \boxed{3.3 \times 10^3\text{ N}}$.

4.29 First make a free body diagram.

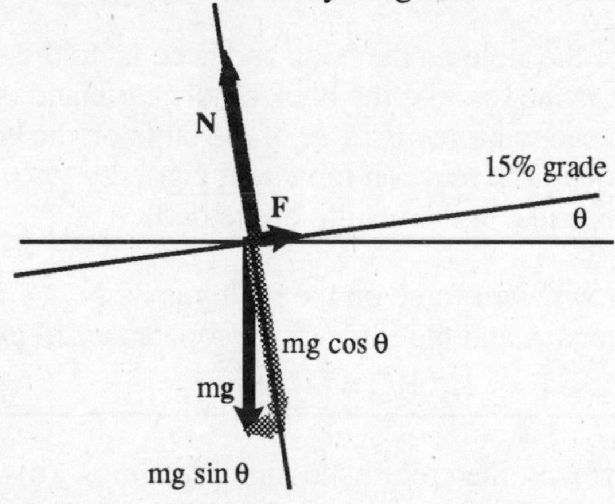

From the diagram we see that the force F upward along the 15% grade must be equal to mg sin θ. So $F = mg \sin\theta$. The angle θ is determined from

$\theta = \tan^{-1}\left(\dfrac{15}{100}\right) = 8.5°$. This gives

$F = (5100\text{ kg})(9.81\text{ m/s}^2)\sin 8.5° = \boxed{7400\text{ N}}$.

4.31 The system of the two masses is accelerated by the net force:

$F_{net} = (m_2 - m_1)g = (m_1 + m_2)a$.

From this we get g in terms of the acceleration.

$$g = \left(\frac{m_1 + m_2}{m_2 - m_1}\right)a.$$

The acceleration is determined from the measurements of distance and time

$$a = \frac{2x}{t^2} = \frac{6.00\text{ m}}{(3.6\text{ s})^2} = 0.46\text{ m/s}^2.$$

$$g = \left(\frac{m_1 + m_2}{m_2 - m_1}\right) a = \left(\frac{2.10}{0.10}\right)(0.46)\ \text{m/s}^2$$

$\boxed{g = 9.7\ \text{m/s}^2.}$

4.33 Friction force = F. In this case the normal force is N = mg. Thus the friction force becomes $F = \mu N = \mu mg = 0.37 \times 50.0\ \text{kg} \times 9.80\ \text{m/s}^2 = \boxed{180\ \text{N}}$.

4.35 Net force along direction of motion is the gravitational force on the 4.00 kg block minus the friction force. The acceleration is the net force divided by the total mass.

$$a = \frac{4.00\ \text{kg} \times 9.80\ \text{m/s}^2 - 0.55 \times 5.00\ \text{kg} \times 9.80\ \text{m/s}^2}{(4.00 + 5.00)\text{kg}},$$

$a = 1.36\ \text{m/s}^2 \approx \boxed{1.4\ \text{m/s}^2}$.

4.37 The relation between distance and acceleration is $s = \frac{1}{2}at^2$.

Use this to calculate the acceleration $a = \frac{2s}{t^2} = \frac{2 \times 20\ \text{m}}{100\ \text{s}^2} = 0.40\ \text{m/s}^2$.

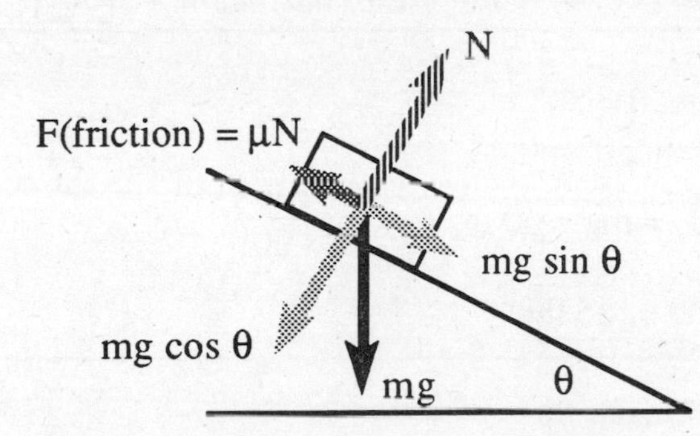

Here, θ = 30°.

From a free body diagram we find $F_{net} = mg \sin\theta - F_{friction} = ma$.

$F_{friction} = m(g \sin\theta - a) = 4.7\ (9.8 \times 0.50 - 0.40)\ \text{N} = \boxed{21\ \text{N}}$.

4.39 (a) The acceleration is given by $a = \frac{\text{net force}}{\text{total mass}}$.

net force = $(5.0\ \text{kg})g - \mu(4.0\ \text{kg})g$, and the total mass = 4.0 kg + 5.0 kg = 9.0 kg.

$$a = \frac{(5.0\ \text{kg} - 0.25 \times 4.0\ \text{kg})9.8\ \text{m/s}^2}{9.0\ \text{kg}} = \frac{4.0}{9.0} 9.8\ \text{m/s}^2 = \boxed{4.4\ \text{m/s}^2}.$$

(b) The net force on the 5.0-kg block is given by $mg - T = ma$. Solving for the tension gives $T = m(g - a) = 5.0\ \text{kg}(9.8 - 4.4)\text{m/s}^2 = \boxed{27\ \text{N}}$.

4.41 The acceleration is the net force divided by the total mass. The net force is the weight of the 6.5-kg block minus the frictional forces.

$$a = \frac{F_{net}}{M_{total}} = \frac{[6.5\ \text{kg} - 0.35(6.0\ \text{kg} + 3.0\ \text{kg})](9.8\ \text{m/s}^2)}{(6.5 + 6.0 + 3.0)\ \text{kg}}$$

$\boxed{a = 2.1\ \text{m/s}^2.}$

$T_2 = (6.5\ \text{kg})(g - a) = 6.5\ \text{kg}(9.8 - 2.1)\text{m/s}^2 = \boxed{50\ \text{N}}$.

$T_1 - \mu(6.0 \text{ kg})g = (6.0 \text{ kg})a$

$T_1 = 6.0 \text{ kg } [2.1 + (0.35)(9.80)] \text{ N} = \boxed{33 \text{ N}}$.

4.43 From the free body diagram, if the paperweight does not slide,
$F_{net} = mg \sin\theta - \mu N = mg \sin\theta - \mu\, mg \cos\theta = 0.$

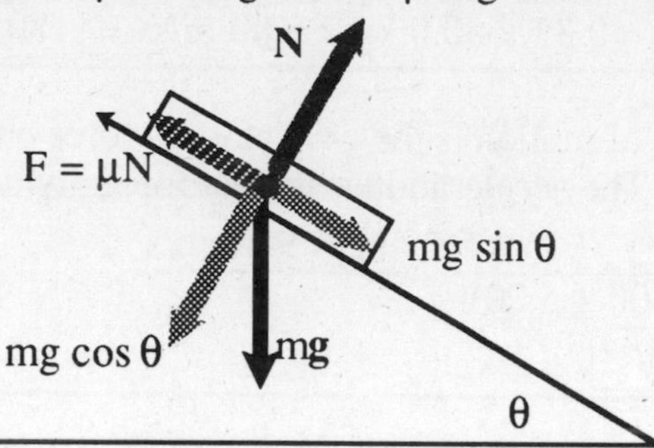

$$\mu = \frac{mg \sin\theta}{mg \cos\theta} = \tan\theta$$

Using $\mu = 0.58$ from table 4.2, we get: $\theta = \tan^{-1}\mu = \tan^{-1}0.58 = \boxed{30°}$.

4.45 Refer to Fig. 4.25 in the text. Here, $\theta = 30°$.
In the horizontal direction, $F - T \sin 30° = 0$.

In the vertical direction, $T \cos 30° = mg = 60 \text{ lb}$; $T = \dfrac{60 \text{ lb}}{\cos 30°}$

$F = 60 \text{ lb} \dfrac{\sin 30°}{\cos 30°} = 60 \text{ lb} \tan 30° = \boxed{35 \text{ lb}}$.

4.47 First make a diagram

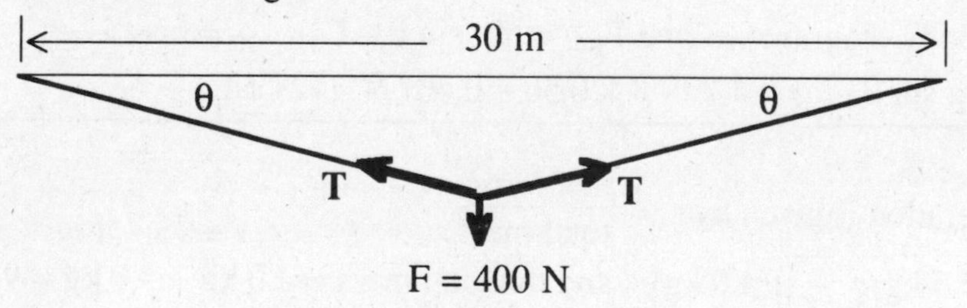

$\tan\theta = \dfrac{\text{displacement}}{\text{one half of distance from car to tree}} = \dfrac{3.0 \text{ m}}{15 \text{ m}}$. $\theta = 11.3°$.

From the vector diagram $2T \sin\theta = 400 \text{ N}$. $T = \dfrac{400}{2 \sin 11.3°} \text{ N}$.

$\boxed{T = 1000 \text{ N.}}$

4.49 For static equilibrium the net horizontal and vertical forces must be zero.
$F_x = T_C \cos 30° - T_A = 0$, and $F_y = T_C \sin 30° - mg = 0$.

$T_A = T_C \cos 30° = mg \dfrac{\cos 30°}{\sin 30°} = \dfrac{mg}{\tan 30°} = \boxed{2.5 \text{ N}}$.

4.51 The sum of the vertical forces is
$F_y = T_1 \sin 30° + T_2 \sin 40° - 20.0\ N = 0$, where up is the positive direction.
The sum of the horizontal forces is
$F_x = T_2 \cos 40° - T_1 \cos 30° = 0$, where we have chosen the positive x direction to be toward the right. Inserting the values for sin and cos,

$0.5\ T_1 + 0.64\ T_2 = 20.0\ N$ and $T_2 = T_1 \frac{0.866}{0.766} = 1.13\ T_1$.

$0.5\ T_1 + 0.64(1.13\ T_1) = 20.0\ N$. $T_1 = \boxed{16.4\ N.}$

$T_2 = 1.13\ T_1 = 1.13(164\ N) = \boxed{18.5\ N}$.

4.53 (a) There is no motion when $m_1 g = m_2 g \sin\theta$. Use this to find θ.

$\theta = \sin^{-1}\frac{m_1}{m_2} = \sin^{-1} 0.5 = \boxed{30°}$.

(b)The acceleration for $\theta = 37°$ is given by $a = \frac{F_{net}}{m_1 + m_2} = \frac{m_2 \sin 37° - m_1}{m_1 + m_2}\ g$,

$a = \frac{10.0 \times 0.60 - 5.0}{15}\ 9.8\ m/s^2 = \boxed{0.65\ m/s^2}$ down the incline.

4.55 Use Newton's second law and the definition of frictional force:

$F = ma = F_{fr} = \mu mg.$ $\mu = \frac{a}{g} = \frac{5.0}{9.8} = \boxed{0.51}$.

4.57 The acceleration is the net force divided by the total mass. The net force is the weight of the 5.0-kg block. $a = \frac{F_{net}}{M_{total}} = \frac{5.0\ kg\ 9.8\ m/s^2}{(5.0 + 10)kg} = \boxed{3.3\ m/s^2}$.

4.59 (a) Force due to car must be equal and opposite to the retarding force if there is no acceleration. $\boxed{F = 100\ N.}$

(b) Difference in forces produces acceleration.

$F - F_{friction} = ma.$ $F - 100\ N = (500\ kg)(2.0\ m/s^2)$

$F = 1000\ N + 100\ N = \boxed{1100\ N}$.

(c) Displacement is related to acceleration and time through $x = \frac{1}{2}at^2$ if $v_o = 0$.

$a = \frac{2x}{t^2}$. $F - F_{friction} = ma = 500\ kg \left(\frac{2 \times 150\ m}{(10\ s)^2}\right) = 1500\ N.$

$F = F_{friction} + 15\ 00\ N = 100\ N + 1500\ N = \boxed{1600\ N}$.

4.61 The net horizontal force is $F_x = F \cos 45° - \mu N = 0$, and the net vertical force is $N = mg - F \sin 45° = 0$. Upon combining these equations we get

$F \cos 45° = \mu[mg - F \sin 45°]$. $\cos 45° = \sin 45° = 1/\sqrt{2}$.

$F/\sqrt{2} = \mu(mg - F/\sqrt{2})$.

$F\left(\frac{1 + \mu}{\sqrt{2}}\right) = \mu\ mg.$ $F = \frac{\sqrt{2}\mu mg}{1+\mu} = \boxed{210\ N}$.

4.63 (a) First we need to compute the acceleration. From v = at we get
$a = v/t = (5.0 \text{ m/s})/(2.0 \text{ s}) = 2.5 \text{ m/s}^2$.
The tension in the cable is found from
$T - Mg = Ma$. Here M is the total mass of elevator plus passengers.
$T = M(a + g) = (2500 \text{ kg} + 260 \text{ kg})(2.5 \text{ m/s}^2 + 9.8 \text{ m/s}^2)$

$\boxed{T = 3.4 \times 10^4 \text{ N.}}$

(b) If the elevator accelerates downward then a must be negative so that the tension is given by $T = M(-a + g) = (2500 \text{ kg} + 260 \text{ kg})(-2.5 \text{ m/s}^2 + 9.8 \text{ m/s}^2)$

$\boxed{T = 2.0 \times 10^4 \text{ N.}}$

4.65 (a) Because the monkey and the bananas have equal masses, the
bananas rise equally with the monkey.

(b)When the monkey traverses a distance d along the rope, the bananas move up d/2 and the monkey also moves up d/2 from its initial position.

$y = \frac{d}{2} = \frac{3 \text{ m}}{2} = \boxed{1.5 \text{ m}}$.

4.67 Refer to the free-body diagrams drawn below.

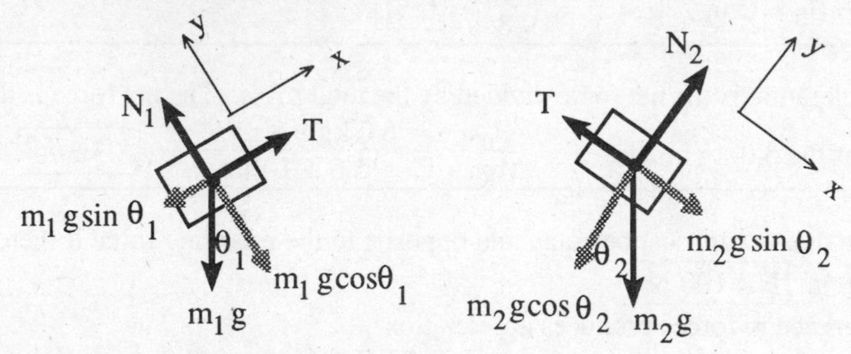

The two force equations are
$T - m_1 g \sin\theta_1 = m_1 a$ and $m_2 g \sin\theta_2 - T = m_2 a$.
Add the equations to eliminate T: $(m_2 \sin\theta_2 - m_1 \sin\theta_1)g = (m_1 + m_2)\,a$.
Rearrange the equation to separate the masses.
$m_2(g \sin\theta_2 - a) = m_1(g \sin\theta_1 + a)$,

$$m_2 = m_1\left(\frac{g \sin\theta_1 + a}{g \sin\theta_2 - a}\right).$$

The acceleration is to the left so $a = -0.010 \text{ m/s}^2$.

$$m_2 = 1.05 \text{ kg}\left(\frac{9.81 \sin 30° - 0.010}{9.81 \sin 40° + 0.010}\right) = 0.814 \text{ kg} \approx \boxed{0.81 \text{ kg}}.$$

Chapter 5

UNIFORM CIRCULAR MOTION AND GRAVITATION

5.1 We can use Eq. (5.3) to relate the speed and the radius to the acceleration.

$a_c = \frac{v^2}{r}$. Upon rearranging we get $v^2 = a_c r$ or $v = \sqrt{a_c r}$.

$v = \sqrt{4300\ \text{m/s}^2 \times 0.0560\ \text{m}} = \boxed{15.5\ \text{m/s.}}$

5.3 Obtain the centripetal acceleration from the frequency. Convert to the proper units.

$$a_c = \omega^2 r = (2\pi f)^2 r = \left[2\pi\left(\frac{100}{3\ \text{min}}\right)\left(\frac{1\ \text{min}}{60\ \text{s}}\right)\right]^2 6.0\ \text{in} \times 2.54\ \text{cm/in} = 186\ \text{cm/s}^2$$

$\boxed{a_c = 1.9\ \text{m/s}^2.}$

5.5 (a) $\omega = 2\pi f = 2\pi\left(\frac{100}{3\ \text{min}}\right)\left(\frac{1\ \text{min}}{60\ \text{s}}\right) = \boxed{3.49\ \text{rad/s}}$.

(b) $\omega = 2\pi f = 2\pi\left(\frac{45}{\text{min}}\right)\left(\frac{1\ \text{min}}{60\ \text{s}}\right) = \boxed{4.71\ \text{rad/s}}$.

5.7 (a) We can find the frequency from $f = \frac{v}{2\pi r}$, but we must be sure to convert the units of the 50 km/h to units of m/s and use r = 0.34 m.

$$f = \frac{(5.0 \times 10^4\ \text{m/h})(1\ \text{h/3600 s})}{2\pi(0.34\ \text{m})} = \boxed{6.5\ \text{Hz}}.$$

(b) $\omega = 2\pi f = \boxed{41\ \text{rad/s}}$.

5.9 (a) Angular velocity $= \omega = \frac{2\pi}{T} = \frac{2\pi}{1.00\ \text{s}} = \boxed{6.28\ \text{rad/s}}$.

(b) $v = \frac{2\pi r}{T} = \frac{2\pi(2.00\ \text{m})}{1.00\ \text{s}} = \boxed{12.6\ \text{m/s}}$.

(c) $a_c = \omega^2 r = \left(\frac{2\pi}{1.00\ \text{s}}\right)^2 (2.00\ \text{m}) = \boxed{79.0\ \text{m/s}^2}$.

5.11 The angular velocity of the wheel is obtained from

$\omega = \frac{\Delta\theta}{\Delta t} = \frac{2.36\ \text{rad}}{0.19\ \text{s}} = 12.4\ \text{rad/s}$. Since $\omega = \frac{v}{r}$, we find that

$r = \frac{v}{\omega} = \frac{2.87\ \text{m/s}}{12.4\ \text{rad/s}} = 0.231\ \text{m}$. Recalling our rules for rounding off we see that we must limit to 2 significant figures so that we get $\boxed{r = 0.23\ \text{m.}}$

5.13 $F_c = m\frac{v^2}{r} = 2000\text{ kg}\frac{(50000\text{ m}/3600\text{ s})^2}{175\text{ m}} = \boxed{2.2 \times 10^3\text{ N}}$.

5.15 The mathematical relationship that we need here is

$$\tan\theta = \frac{v^2}{rg} = \frac{(8.0 \times 10^4\text{ m}/3600\text{ s})^2}{200\text{ m} \times 9.80\text{ m/s}^2} = 0.252.$$

$\theta = \tan^{-1} 0.252 = \boxed{14°}$.

5.17 First find the horizontal component of the force, the centripetal force:

$F_c = m\omega^2 r = m4\pi^2 f^2 r = (0.208\text{ kg})(4\pi^2)(3.00\text{ hz})^2(1.00\text{ m}) = 73.9\text{ N}$

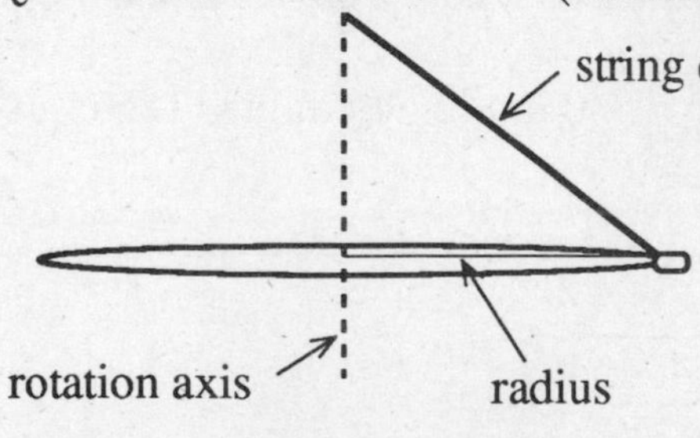

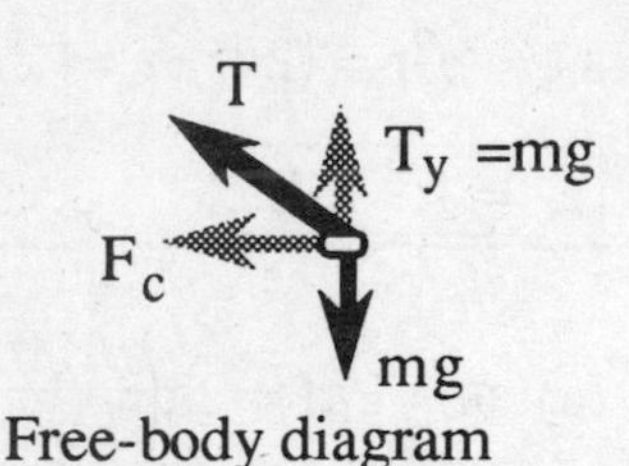

Free-body diagram

The total force T of the string includes a horizontal component necessary for circular motion and a vertical component to oppose the force of gravity. So that the maximum force T becomes

$F = \sqrt{F_c^2 + (mg)^2} = \sqrt{(73.9)^2 + (1.96)^2}\text{ N} = \boxed{73.9\text{ N}}$.

5.19 (a) Force due to the basket at the top becomes zero when the required centripetal acceleration is provided by gravity. Thus $a_c = \omega^2 r = g$,

$$f = \frac{\omega}{2\pi} = \frac{1}{2\pi}\sqrt{\frac{g}{r}} = \frac{1}{2\pi}\sqrt{\frac{9.81\text{ m/s}^2}{0.65\text{ m}}} = \boxed{0.62\text{ Hz}}.$$

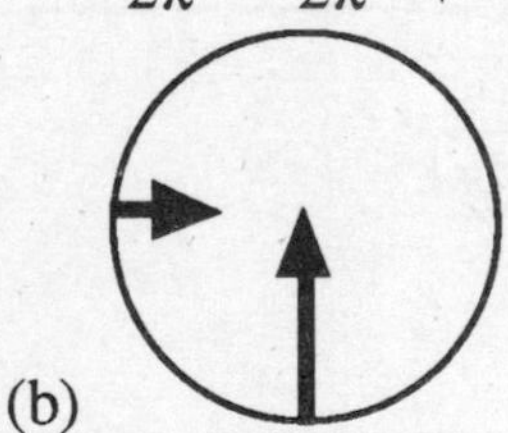

(b)

5.21 The centripetal force may be expressed by $F_c = \frac{mv^2}{r}$. Here, F_c and v are given as the same for both particles.

$\frac{m_e v^2}{r_e} = \frac{m_p v^2}{r_p}$, so

$$r_p = r_e\frac{m_p}{m_e} = 2.85\text{ cm}\frac{1.67 \times 10^{-27}\text{ kg}}{9.11 \times 10^{-31}\text{ kg}} = 5.22 \times 10^3\text{ cm} = \boxed{52.2\text{ m}}.$$

5.23 Make a vector diagram of all forces.

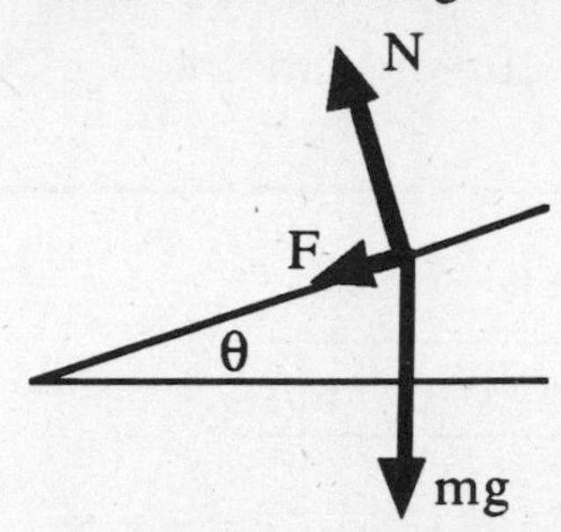

(1) In the vertical direction we see that $N\cos\theta = F\sin\theta + mg$. (2) In the horizontal direction we have $N\sin\theta + F\cos\theta = mv^2/r$, the centripetal force. At the design speed of 120 km/h, the force parallel to the road is $F = 0$. Then, $N_o\cos\theta = mg$ and $N_o\sin\theta = mv_o^2/r$, so $\tan\theta = v_o^2/gr$.
Now we can insert the values for v_o and r to get $\tan\theta = 0.151$, or $\theta = 8.596°$.
To find the frictional force F at other speeds, divide Eq.(2) by Eq.(1) to get

$$\tan\theta = v_o^2/r = \frac{mv^2/r - F\cos\theta}{mg + F\sin\theta}.$$ This equation can be rearranged as

$$F = \frac{\frac{m}{r}(v^2 - v_o^2)}{\sin\theta\tan\theta + \cos\theta} = \boxed{6.23\times10^3\ \text{N}}.$$

5.25 Use the results of Example 5.10: $g_{moon} = \frac{GM_m}{r_m^2}$ and $g_{earth} = \frac{GM_E}{r_E^2}$

$$\frac{g_{moon}}{g_{earth}} = \left(\frac{M_m}{M_E}\right)\left(\frac{r_E}{r_m}\right)^2 = (0.0123)\left(\frac{1}{0.273}\right)^2 = \boxed{0.165}.$$

5.27 The change in weight is due to a change in local gravitational acceleration g.

$\Delta W = m\,\Delta g = m\,GM_E\left(\frac{1}{R_E^2} - \frac{1}{(R_E + h)^2}\right)$, where R_E is the earth's radius and h is the height of the mountain.

$$\Delta W = \frac{GmM_E}{R_E^2}\left[1 - \left(1 + \frac{h}{R_E}\right)^{-2}\right],$$

$$\Delta W \approx W\left[1 - \left(1 - \frac{2h}{R_E}\right)\right] = -\frac{2h}{R_E}W,$$

$$\Delta W \approx -\frac{2\,(2.05\text{ km})}{6.37\times10^3\text{ km}}(2.10\times10^3\text{ kg})(9.81\text{ m/s}^2) = \boxed{-13.3\text{ N}}.$$

5.29 (a) Density is defined as ρ = mass/volume. So

$$V = \frac{m}{\rho} = \frac{0.500\text{ kg}}{1.93\times10^4\text{ kg/m}^3} = \boxed{2.59\times10^{-5}\text{ m}^3}.$$

(b) For a cube of length L the volume is $V = L^3$, so

$$L = V^{1/3} = 0.0296\text{ m} = \boxed{2.96\text{ cm}}.$$

5.31 Density is defined as ρ = mass/volume = m/V. So

$m = \rho V = 1.13 \times 10^4\ \text{kg/m}^3 \times 5.0 \times 10 \times 30\ \text{cm}^3 \times 10^{-6}\ \text{m}^3/\text{cm}^3$

$m = 16.95\ \text{kg} \approx \boxed{17\ \text{kg}}$.

5.33 (a) In an inertial frame, the acceleration of the moon is

$$a_m = \frac{GM_E}{R^2} = \frac{(6.67 \times 10^{-11})(5.98 \times 10^{24})}{(3.84 \times 10^8)^2} = \boxed{2.70 \times 10^{-3}\ \text{m/s}^2}.$$

(b) The acceleration of the earth is

$$a_m = \frac{GM_m}{R^2} = \frac{(6.67 \times 10^{-11})(7.36 \times 10^{22})}{(3.84 \times 10^8)^2} = \boxed{3.33 \times 10^{-5}\ \text{m/s}^2}.$$

5.35 $$F = \frac{G\,M\,m}{r^2}.$$

$$F = \frac{(6.67 \times 10^{-11}\ \text{N m}^2/\text{kg}^2)(1.99 \times 10^{30}\ \text{kg})(14.5 \times 5.98 \times 10^{24}\ \text{kg})}{(19.2 \times 1.5 \times 10^{11}\ \text{m})^2}.$$

$$F = \boxed{1.39 \times 10^{21}\ \text{N}}.$$

5.37 Equate the gravitational force to the centripetal force expressed in terms of the period to get $M_J = \frac{4\pi^2 r^3}{GT^2}$.

$$M_J = \frac{4\pi^2(4.22 \times 10^8\ \text{m})^3}{6.67 \times 10^{-11}\ \text{N}\cdot\text{m}^2/\text{kg}^2(42.5 \times 3600\ \text{s})^2} = \boxed{1.90 \times 10^{27}\ \text{kg}}.$$

5.39 The period is T = 88 min. Equate the gravitational force to the centripetal force expressed in terms of the period to get $r^3 = \frac{GM_eT^2}{4\pi^2}$, $r = 6.56 \times 10^6$ m.

$$r - R_e = (6.56 - 6.38) \times 10^6\ \text{m} = 0.18 \times 10^6\ \text{m} = \boxed{180\ \text{km}}.$$

5.41 (a) Equate the gravitational force to the centripetal force expressed in terms of the period to get $M_s = \frac{4\pi^2 r_E^3}{GT_E^2}$

$$M_s = \frac{4\pi^2(1.5 \times 10^{11}\ \text{m})^3}{6.67 \times 10^{-11}\ \text{N}\cdot\text{m}^2/\text{kg}^2(365 \times 86400\ \text{s})^2} = \boxed{2.0 \times 10^{30}\ \text{kg}}.$$

(b) $$\rho = \frac{M_s}{V} = \frac{M_s}{\frac{4}{3}\pi(6.96 \times 10^8\text{m})^3} = \boxed{1400\ \text{kg/m}^3}.$$

Compared to the earth's density of 5.5×10^3 kg/m^3

$$\frac{\rho_s}{\rho_E} = \frac{1400}{5500} = \boxed{0.25}.$$

5.43 From the law of universal gravity we get "g" $= \frac{GM_E}{(R_E+h)^2}$, where h is the height above the earth. As a ratio to g at the earth's surface we get

$$\frac{\text{"g"}}{g} = \frac{R_E^2}{(R_E+h)^2} = \frac{1}{(1 + h/R_E)^2} \approx 1 - 2h/R_E = 1 - \frac{2(600 \text{ km})}{6380 \text{ km}} = 0.812.$$

Multiply by 100 to get percent: $\frac{\text{"g"}}{g} = \boxed{81.2\%.}$

5.45 Weight on Mars $= \frac{GmM_M}{R_M^2} = m\frac{G(0.107)M_E}{(0.53)^2R_E^2} = mg\frac{0.107}{(0.53)^2}$

Weight on Mars $= 700 \text{ N}\frac{0.107}{(0.53)^2} = \boxed{267 \text{ N}}$.

5.47 First calculate the acceleration on Mars.

$$a = \frac{GM_m}{(R_m)^2} = \frac{G(0.107)M_e}{(0.530 R_m)^2} = 0.381 \text{ g}.$$

$$y = \frac{1}{2}at^2 = \frac{1}{2}(0.381)(9.81 \text{ m/s}^2)(1.00 \text{ s})^2 = \boxed{1.87 \text{ m}}.$$

5.49 $\Gamma = \frac{GM_E}{R_E^2} = \frac{(6.67 \times 10^{-11} \text{ N·m}^2/\text{kg}^2)(5.98 \times 10^{24} \text{ kg})}{(6.38 \times 10^6 \text{ m})^2} = \boxed{9.80 \text{ m/s}^2}$.

Direction is radial toward the center of the earth.

5.51 First make a diagram of the situation.

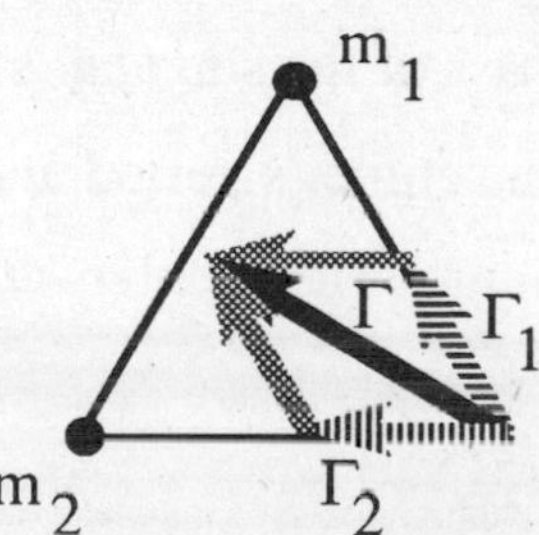

The direction of the vector sum Γ is toward the midpoint of the line joining the masses.

$$\left|\Gamma\right| = 2 \cos 30° \frac{Gm}{r^2} = 2 \cos 30° \frac{(6.67 \times 10^{-11} \text{ N·m}^2/\text{kg}^2)(1.0 \text{ kg})}{(1.0 \text{ m})^2}$$

$$\left|\Gamma\right| = \boxed{1.2 \times 10^{-10} \text{ m/s}^2}.$$

5.53 Density is defined as ρ = mass/volume = m/V. So

$$\rho = \frac{m}{V} = \frac{4.45 \text{ kg}}{5.0 \times 5.0 \times 20 \times 10^{-6} \text{ m}^3} = \boxed{8.9 \times 10^3 \text{ kg/m}^3}.$$

5.55 (a) $v = r\omega = 2\pi rf$ is the same for both, so

$$f_s = \frac{r_1 f_1}{r_s} = \left(\frac{15\text{ cm}}{5\text{ cm}}\right) 10\text{ Hz} = \boxed{30\text{ Hz}}.$$

(b) $v = 2\pi rf = 2\pi(15\text{ cm})(10\text{ Hz}) = 940\text{ cm/s} = \boxed{9.4\text{ m/s}}.$

5.57 Use the law of universal gravitation to find a force 10^{-3} times your weight mg.

$$\frac{GMm}{r^2} = 10^{-3}\text{ mg} \quad \text{or} \quad \frac{GM}{r^2} = 10^{-3}\text{ g}.$$

But $M = \rho V = \rho \frac{4}{3}\pi r^3$, so

$$G\tfrac{4}{3}\rho\pi r = 10^{-3}\text{ g}$$

$$r = \frac{3 \times 10^{-3}\text{ g}}{4G\rho\pi} = \frac{(3 \times 10^{-3})(9.81\text{ m/s}^2)}{4(6.67 \times 10^{-11}\text{ N}\cdot\text{m}^2/\text{kg}^2)(1.13 \times 10^4\text{ kg/m}^3)\pi}$$

$$r = \boxed{3100\text{ m}}.$$

5.59 The centripetal acceleration needed at the top must equal or exceed the gravitational acceleration: $g = \omega^2 r$.

$$f = \frac{\omega}{2\pi} = \frac{1}{2\pi}\sqrt{\frac{g}{r}} = \frac{1}{2\pi}\sqrt{\frac{9.80\text{ m/s}^2}{4.5\text{ m}}} = \boxed{0.23\text{ Hz}}.$$

5.61 From the force law or Kepler's third law we get

$\frac{T^2}{R^3} = \frac{4\pi}{GM}$. The constant on the right contains the mass of the sun and is the same for the earth and for the comet. So we can compare the comet's period and distance to that of the earth by $\frac{T_c^2}{R_c^3} = \frac{T_E^2}{R_E^3}$.

If we use time in years and distance in AU we get

$$R_c^3 = R_E^3 \frac{T_c^2}{T_E^2} \quad \text{or} \quad R_c = R_E\left(\frac{Tc2}{TE^2}\right)^{1/3} = 1.0\text{ AU}\left(\frac{75^2}{1.0^2}\right)^{1/3}$$

$R_c = 17.8$ AU. Since the average distance is the semimajor axis of the ellipse, the full major axis must be a length $2R_c = 35.6\text{ AU} \approx 36$ AU. If at the nearest point the comet passes much less than an AU from the sun, the farthest distance must be nearly $2R_c \approx \boxed{36\text{ AU.}}$

5.63 (a) From the force law (or Kepler's third law) we get

$\frac{T^2}{R^3} = \frac{4\pi^2}{GM}$, where is R is the earth-moon distance. So $M = \frac{4\pi^2 R^3}{GT^2}$,

$$M = \frac{4\pi^2(3.84 \times 10^8\text{ m})}{6.67 \times 10^{-11}\text{ N}\cdot\text{m}^2/\text{kg}^2(27.3\cdot 24\cdot 3600\text{ s})^2},$$

$M = \boxed{6.02 \times 10^{24} \text{ kg}}$.

(b) $\rho = \frac{M}{V} = \frac{M}{\frac{4}{3}\pi R_E{}^3} = \boxed{5.53 \times 10^3 \text{ kg/m}^3}$.

5.65 Maximum tension of 2 strings is $T = 2(Mg)$. In circular motion about a vertical circle the centripetal force when the weight is at the bottom is $F = M(2\pi f)^2 r$. The tension in the string must supply the centripetal force plus support the weight of the mass. So $T = F + Mg = 2(Mg)$.

$F = Mg$, and $(2\pi f)^2 r = g$. Thus $f = \frac{1}{2\pi}\sqrt{\frac{g}{r}} = \frac{1}{2\pi}\sqrt{\frac{9.80}{0.75}}$ Hz $= \boxed{0.57 \text{ Hz}}$.

5.67 (a) $\frac{F}{m} = \text{“g”} = \frac{GM_E}{R_{EM}{}^2}\left(\frac{1}{\left(\frac{1}{4}\right)^2} - \frac{0.0123}{\left(\frac{3}{4}\right)^2}\right)$,

$$\text{“g”} = \frac{6.67 \times 10^{-11} \times 5.98 \times 10^{24}}{(3.84 \times 10^8)^2}\left(16 - \frac{16}{9} \times 0.0123\right) \text{m/s}^2,$$

$$\text{“g”} = 2.70 \times 10^{-3} \times 15.98 \text{ m/s}^2 \times \frac{g}{9.81 \text{ m/s}^2} = \boxed{4.4 \times 10^{-3}\, g}.$$

(b) $\text{“g”} = 2.70 \times 10^{-3}\,(4 - 4(0.0123)) \text{ m/s}^2 \times \frac{g}{9.81 \text{ m/s}^2} = \boxed{1.1 \times 10^{-3}\, g}$.

(c) $\text{“g”} = 2.70 \times 10^{-3}\left(\frac{16}{9} - 16(0.0123)\right) \text{m/s}^2 \times \frac{g}{9.81 \text{ m/s}^2} = \boxed{4.3 \times 10^{-4}\, g}$.

5.69 From Eq.(5.10) we relate the bank angle to speed and gravity through $\tan\theta = v^2/gr$. We assume same speed and radius, but different g on the moon and find the new angle. First, on earth we have

$$\tan\theta_E = \frac{\left(240 \text{ km/h} \times \frac{1000 \text{ m/km}}{3600 \text{ s/h}}\right)^2}{(9.81 \text{ m/s}^2)(1500 \text{ m})} = 0.302, \text{ so } \theta_E = 16.8°.$$

On the moon we get

$$\tan\theta_m = \frac{\left(240 \text{ km/h} \times \frac{1000 \text{ m/km}}{3600 \text{ s/h}}\right)^2}{(1.62 \text{ m/s}^2)(1500 \text{ m})} = 1.83, \text{ so } \theta_m = 61.3°.$$

The angle will have to be increased by the difference 61.3°-16.8° = 44.5°.
The angle must be $\boxed{\text{increased by } 44.5°}$.

5.71 From the force law (or Kepler's third law) we get

$\frac{T^2}{r^3} = \frac{4\pi^2}{GM}$, where r is the planet's radius = radius of the satellite's orbit, T is the period of the orbit, and M is the mass of the planet. Knowing the period and the speed of the satellite, we can calculate it's orbit radius. Then using that as the radius of the planet, we can compute the mass of the planet from it volume and density.

The speed of the satellite is $v = \frac{2\pi r}{T}$, so $r = \frac{vT}{2\pi}$.

From the Kepler law we substitute for $M = \frac{4}{3}\rho\,\pi r^3$ to get

$$T = \sqrt{\frac{4\pi^2 r^3}{G\frac{4}{3}\rho\,\pi r^3}} = \sqrt{\frac{3\,\pi}{\rho\,G}}.$$

Then $r = \frac{vT}{2\pi} = \frac{v}{2\pi}\sqrt{\frac{3\,\pi}{\rho\,G}}$.

The mass is then found from

$$M = \frac{4}{3}\rho\,\pi r^3 = \frac{4}{3}\rho\,\pi\left(\frac{v}{2\pi}\right)^3\left(\frac{3\pi}{\rho G}\right)^{3/2} = \left(\frac{v^3}{2\pi G}\right)\sqrt{\frac{3\pi}{\rho G}},$$

$$M = \left(\frac{(3.55\times10^3\ \text{m/s})^3}{2\pi\times6.67\times10^{-11}\ \text{m}^3/\text{kg}\cdot\text{s}^2}\right)\times$$

$$\sqrt{\frac{3\pi}{(3.90\times10^3\ \text{kg/m}^3)(6.67\times10^{-11}\ \text{m}^3/\text{kg}\cdot\text{s}^2)}},$$

$M = \boxed{6.42\times10^{23}\ \text{kg}}$, or about 0.1 time the earth's mass.

Chapter 6

WORK AND ENERGY

6.1 $W = Fx = mgh = (55 \text{ kg})(9.81 \text{ m/s}^2)(3.0 \text{ m}) = \boxed{1.6 \times 10^3 \text{ J}}$.

6.3 The work is done by the horizontal component of the force.

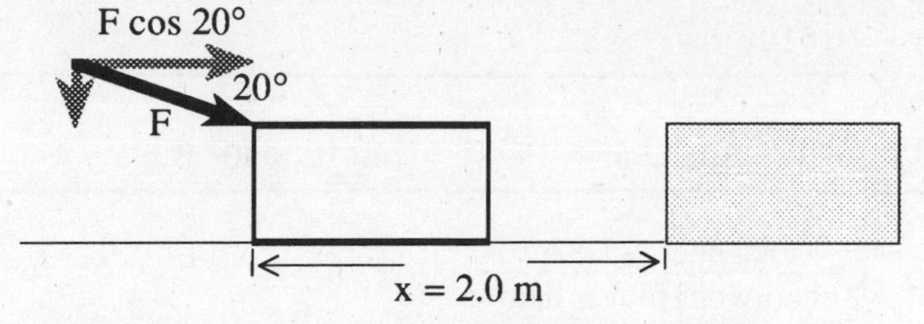

$W = Fx \cos\theta = (30 \text{ N})(2.0 \text{ m})(\cos 20°) = \boxed{56 \text{ J}}$.

6.5 The force is proportional to the extension: $F = kx$. Thus we have $k = F/x$. The energy stored is given by $W = \frac{1}{2}kx^2$. Substituting for k we get

$W = \frac{1}{2}Fx = \frac{1}{2}(20 \text{ N})(5.0 \text{ cm})(10^{-2} \text{ m/cm}) = \boxed{0.50 \text{ J}}$.

6.7 The work done is $W = \frac{1}{2}kx^2$. We can get the spring constant in terms of the work as $k = 2W/x^2$. The force required to stretch the spring a distance x is given by $F = kx = (2W/x^2)x = 2W/x = (2 \times 15 \text{ J})/0.10 \text{ m} = \boxed{300 \text{ N}}$.

6.9 The force up the plane must equal $mg \sin\theta$.

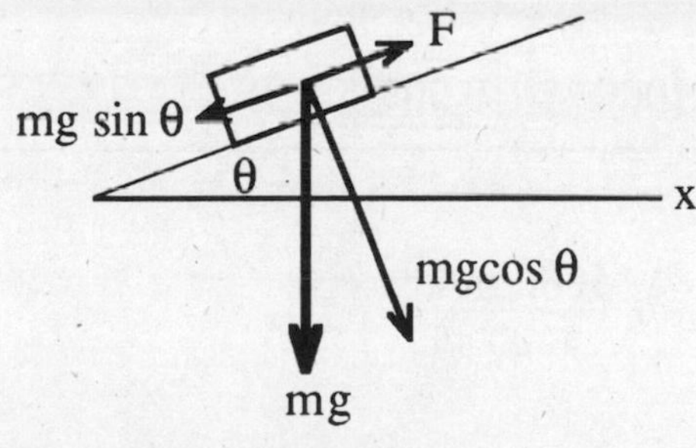

The work is $W = Fd$.

$W = (mg \sin\theta)(d) = (125 \text{ kg})(9.81 \text{ m/s}^2)(\sin 20°)(2.75 \text{ m})$

$\boxed{W = 1.15 \times 10^3 \text{ J.}}$

6.11 If the spring obeys Hooke's law then the energy to stretch it is given by $\frac{1}{2}kx^2$. So the ratio of energies should equal the ratio of the squares of the extensions.

Ratio of energies is $\frac{270 \text{ J}}{46 \text{ J}} = 5.87$.

Ratio of squares of the extensions is $\frac{27^2}{12^2} = 5.06$. Since these ratios are not the same, the spring does not obey Hooke's law. The answer is $\boxed{\text{no.}}$

6.13 (a) From Eq. (6.3) we get the energy stored as

$W = \frac{1}{2}kx^2$. But the force of the gymnast on the spring is mg = kx. So

$W = \frac{1}{2}mgx = \frac{1}{2}(50\text{ kg})(9.81\text{ m/s}^2)(0.50\text{ m}) = \boxed{120\text{ J}}$.

(b) $W = \frac{1}{2}mgx_2 = \boxed{61\text{ J}}$.

6.15 $1\text{ kWh} = 3.600 \times 10^6\text{ J }(1\text{ Btu}/1.055 \times 10^3\text{ J}) = \boxed{3.412 \times 10^3\text{ Btu}}$.

6.17 $\text{Number of plants} = \frac{\text{total energy USA}}{\text{energy of Hoover dam}}$

$\text{number} = \frac{8.6 \times 10^{19}\text{ J}}{2 \times 10^{16}\text{ J}} = \boxed{4300}$.

6.19 (a) Total power $P = \frac{\text{power}}{\text{unit area}} \times \text{area}$,

$P = 180\text{ W/m}^2 \times 3.5 \times 10^6\text{ mi}^2 \times (1609\text{ m/mi})^2 = \boxed{1.6 \times 10^{15}\text{ W}}$.

(b) First compute the total power needed as the per capita power use times the population. Then the area needed at 100% efficiency is the total power divided by the power per unit area. Finally, divide that result by the efficiency.

$\text{Area} = \frac{\text{(power/capita)(population)}}{0.10\text{(power/area)}}$,

$\text{area} = \frac{1.3 \times 10^3\text{ W} \times 2.5 \times 10^8}{0.1 \times 180\text{ W/m}^2} = \boxed{1.8 \times 10^{10}\text{ m}^2}$.

(c) $\%\text{ area covered} = \frac{\text{covered area}}{\text{total area}} \times 100\% = \boxed{0.20\%}$.

6.21 From the definition of kinetic energy:

$KE = \frac{1}{2}mv^2 = \frac{1}{2}(5.0 \times 10^5\text{ kg})\left(90\text{ km/h}\frac{1000\text{ m/km}}{3600\text{ s/h}}\right)^2$

$KE = \boxed{1.6 \times 10^8\text{ J}}$.

6.23 $\frac{KE\text{ (80 mph)}}{KE\text{ (20 mph)}} = \frac{v_2^2}{v_1^2} = \left(\frac{80}{20}\right)^2 = \boxed{16}$.

6.25 Since the mass of the baseball is the same for each, the ratio of the kinetic energies is

$\frac{KE_{Sam}}{KE_{Bill}} = \frac{(v_{Sam})^2}{(v_{Bill})^2} = \boxed{4}$.

6.27 (a) Use the work-energy theorem. $W = \frac{1}{2}mv_f^2 - \frac{1}{2}mv_i^2 = \frac{1}{2}m(v_f^2 - v_i^2)$,

$$W = \frac{1}{2}(1200\ \text{kg})\left[(30\ \text{km/h})^2 - (10\ \text{km/h})^2\right]\left(\frac{1000\ \text{m/km}}{3600\ \text{s/h}}\right)^2.$$

The last term is a conversion factor to change the units. The work becomes

$W = \boxed{3.7 \times 10^4\ \text{J}}$.

(b) $W = \frac{1}{2}(1200\ \text{kg})\left[(50\ \text{km/h})^2 - (30\ \text{km/h})^2\right]\left(\frac{1000\ \text{m/km}}{3600\ \text{s/h}}\right)^2$,

$W = \boxed{7.4 \times 10^4\ \text{J}}$.

6.29 (a) $PE = mgh = (10.0\ \text{kg})(9.81\ \text{m/s}^2)(2.00\ \text{m}) = \boxed{196\ \text{J}}$.

(b) $PE = 196\ \text{J} \times \frac{1\ \text{ft-lb}}{1.356\ \text{J}} = \boxed{145\ \text{ft·lb}}$.

6.31 (a) The potential energy depends on the height above the table.

$PE = mgh = 2.5\ \text{kg}\ (9.81\ \text{m/s}^2)(0.75\ \text{m}) = \boxed{18\ \text{J}}$.

(b) The potential energy depends on the height above the floor which is the height above the table plus the height of the table above the floor.

$PE = mg(h_1 + h_2) = 2.5\ \text{kg}\ (9.81\ \text{m/s}^2)(0.75\ \text{m} + 0.80\ \text{m}) = \boxed{38\ \text{J}}$.

6.33 There is no work done during the purely horizontal movements.

$W_1 = 1.00\ \text{kg}\ (9.81\ \text{m/s}^2)\ 1.00\ \text{m} = \boxed{9.81\ \text{J}}$.

$W_2 = 1.00\ \text{kg}\ (9.81\ \text{m/s}^2 \times \cos 45°)\ (1.00\ \text{m}/\cos 45°) = \boxed{9.81\ \text{J}}$.

$W_3 = 1.00\ \text{kg}\ (9.81\ \text{m/s}^2)\ 1.00\ \text{m} = \boxed{9.81\ \text{J}}$.

$W_4 = 1.00\ \text{kg}\ (9.81\ \text{m/s}^2)\ (1.50\ \text{m} - 0.50\ \text{m}) = \boxed{9.81\ \text{J}}$.

6.35 Use the form of Eq. (6.8): $PE = -\frac{GM_Em}{R_E}$,

$$PE = -\frac{(6.67 \times 10^{-11}\ \text{N·m}^2/\text{kg}^2)(5.98 \times 10^{24}\ \text{kg})(1.00\ \text{kg})}{6.38 \times 10^6\ \text{m}},$$

$PE = \boxed{-6.25 \times 10^7\ \text{J}}$.

6.37 Use the form of Eq. (6.8):

$$\Delta PE = PE_{final} - PE_{initial} = -GM_Em\left(\frac{1}{2R_E} - \frac{1}{R_E}\right) = \frac{GM_Em}{2R_E},$$

$\Delta PE = \boxed{3.12 \times 10^7\ \text{J}}$.

6.39 Use conservation of energy: initial potential energy = final kinetic energy.

$\frac{1}{2}mv^2 = mgh$.

$v = \sqrt{2gh} = \sqrt{2(9.81\ \text{m/s}^2)\ 18\ \text{m}} = \boxed{19\ \text{m/s}}$.

6.41 The initial kinetic energy is zero. Since no energy is lost to friction, the final kinetic energy is equal to the change in potential energy.

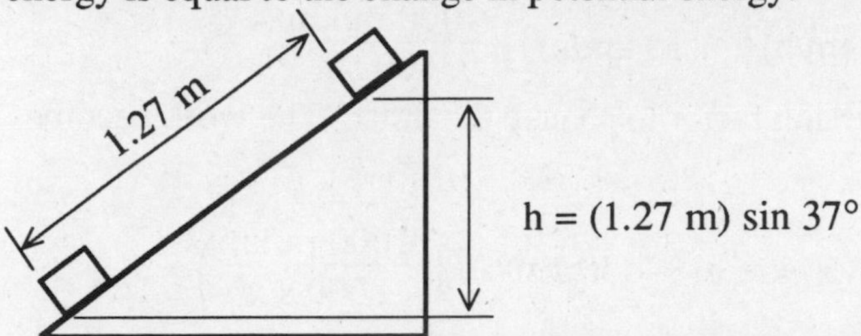

The vertical travel of the block is h = (1.27 m) sin37°. The energy equation is

$\frac{1}{2}mv^2 = mgh,$

$v = \sqrt{2gh} = \sqrt{2(9.81\ m/s^2)(1.27\ m\ \sin 37°)} = \boxed{3.87\ m/s}.$

6.43 At B, height is R – R cos 45°. Use conservation of energy.

PE + KE = $PE_o + KE_o$. At A, the initial kinetic energy is $KE_o = 0$.

$mg(R - R\cos 45°) + \frac{1}{2}mv^2 = mgR.$

$v^2 = 2gR\cos 45° = 2(9.8\ m/s^2)(1\ m)(0.707)$

$v = \boxed{3.7\ m/s}.$

6.45 Equate the potential energy at height h to the kinetic energy at 60 km/h. First find the speed in SI units: $60\ km/h \times \frac{1000\ m/km}{3600\ s/h} = 16.7\ m/s.$

$mgh = \frac{1}{2}mv^2.$ So $h = \frac{v^2}{2g} = \frac{(16.7\ m/s)^2}{2(9.80\ m/s^2)} = \boxed{14\ m}.$

6.47 The energy at the height from which the hammer is dropped is the same as the energy at 0.32 m above the surface on which the astronaut stands. Since the initial kinetic energy is zero, conservation of total mechanical energy gives

$mgh_1 = \frac{1}{2}mv_2^2 + mgh_2$, where g is the local acceleration of gravity.

Rearranging to solve for g and inserting the numerical values gives

$$g = \frac{v_2^2}{2(h_1 - h_2)} = \frac{(4.1\ m/s)^2}{2(1.47\ m - 0.32\ m)} = 7.3\ m/s^2$$

This value is not within the range of values of g found on earth and therefore the answer is $\boxed{\text{no,}}$ the astronaut is not on earth.

6.49 Find the speed at point B to make $a_c = v^2/R = 2g$. Use conservation of energy:

$E = PE_{(A)} = PE_{(B)} + KE_{(B)}$

$mgh = mg(2R) + \frac{1}{2}m(2gR)$

$h = 2R + R = \boxed{3R}.$

6.51 $E = KE + PE = \frac{1}{2}mv^2 - \frac{GM_Em}{R_E}.$

For simplicity, we work this as if the earth is fixed so that the moon moves in a circular orbit. The centripetal force is supplied by gravitation so

$\frac{mv^2}{R_{Em}} = \frac{GM_Em}{(R_{Em})^2}$, which simplifies to $v^2 = \frac{GM_E}{R_{Em}}$. Insert this value of v^2 into the equation for total energy E to get

$$E = -\frac{GM_{Em}m}{2R_E},$$

$$E = -\frac{(6.67 \times 10^{-11}\ Nm^2/kg^2)(5.98 \times 10^{24}\ kg)(7.36 \times 10^{22}\ kg)}{2(3.84 \times 10^8\ m)},$$

$$E = \boxed{-3.82 \times 10^{28}\ J}.$$

6.53 Use conservation of energy. At B: $mgh_A = mgh_B + \frac{1}{2}mv^2$.

$v = \sqrt{2g(h_A - h_B)} = \sqrt{2(9.8)(10-3)}\ m/s = \boxed{11.7\ m/s}$.

At C: $v = \sqrt{2g(h_A - h_C)} = \boxed{8.9\ m/s}$.

At D: $v = \sqrt{2g(h_A - h_D)} = \boxed{9.9\ m/s}$.

To stop at E: $W = Fx = mg(h_A - h_C)$

$a = \frac{F}{m} = \frac{g(h_A - h_C)}{x} = \frac{9.8\ m/s^2\ (10-5)m}{30\ m} = \boxed{1.6\ m/s^2}$.

The direction of a is opposite to the direction of motion.

6.55 $E_{initial} = E_{final} + W_{friction}$

$mgh = \frac{1}{2}mv^2 + 0.10\ mgh$

$\frac{1}{2}v^2 = gh(1 - 0.10)$

$v = \sqrt{2\ gh(0.90)} = \boxed{5.9\ m/s}$.

6.57 First find the coefficient of friction. Make a diagram.

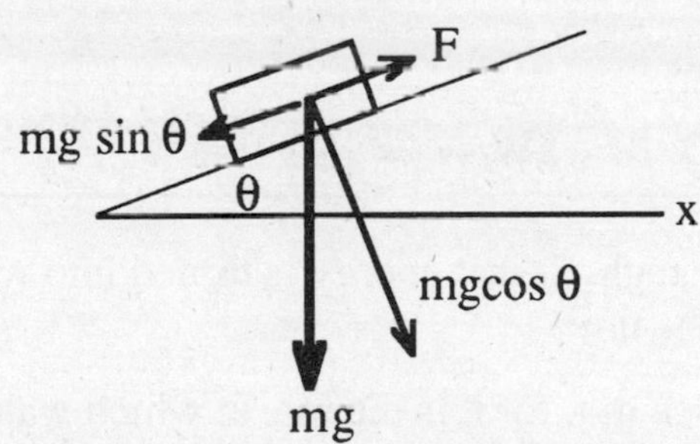

The friction force $F_{fr} = \mu mg\cos\theta$. When the angle $\theta = 34°$, the magnitude of the friction force equals the gravitational component $mg \sin\theta = mg \sin 34°$. So $\mu = \tan 34° = 0.67$.

(a) This part can be done knowing the change in height.

$\Delta PE = PE_{final} - PE_{initial} = mgh_{final} - mgh_{initial} = mg\Delta h$.

$\Delta h = -L \sin 34°$, where L is the length of the board.

So $\Delta PE = mg\Delta h = \boxed{-mgL \sin 34°}$.

(b) Work done by gravity against friction is $F_{fr}L = \mu mg \cos 34°\ L$. But we showed above that

$\mu = \tan 34°$. So $F_{fr}L = \tan 34°\ mg \cos 34°\ L = \boxed{mgL \sin 34°}$.

(c) There is no increase in kinetic energy because the speed was constant. The potential energy lost went into work done against the force of friction. From conservation of energy we get $E_{initial} = E_{final} + W_{friction}$. Rearranging we find $E_{initial} - E_{final} = -\Delta PE = W_{friction}$. This result is in agreement with the results computed in parts (a) and (b).

6.59 Use the conversion factor of 746 W/hp. $\dfrac{250\ W}{746\ W/hp} = \boxed{0.335\ hp}$.

6.61 First convert the 80 hp to units of watts. Then set that power equal to the power of N batteries at 300 W/battery.

80 hp × 746 W/hp = 300 W/battery × N batteries

$$N = \frac{80 \times 746}{300} = \boxed{199}.$$

6.63 The work done in climbing the hill is W = m"g"h where "g" is to be found. The power is the work divided by the time.

$$P = \frac{m\text{"g"}h}{t}. \quad \text{So "g"} = \frac{Pt}{mh} = \frac{200\ W \times 7.2\ s}{110\ kg \times 7.3\ m} = 1.79\ m/s^2.$$

$\boxed{\text{No}}$, the astronaut is not on the earth.

6.65 Power is work divided by time: $P = \dfrac{Fx}{t}$. The power is P = 1.0 hp = 746 W.

Rearrange to get

$$t = \frac{Fx}{P} = \frac{mgx}{P} = \frac{(25\ kg)(9.81\ m/s^2)(10\ m)}{746\ W} = \boxed{3.3\ s}.$$

6.67 Cost = power × time × cost/unit power.

Cost = 250 W × 12 h × \$0.083/kWh × 10^{-3} kW/W = $\boxed{\$0.25}$.

6.69 The available potential energy is PE = mgh. It that energy is turned into work, then the power is the work divided by the time.

$P = \dfrac{PE}{t} = \dfrac{mgh}{t} = \dfrac{m}{t}gh$. We are given m/t, for it is the rate at which water passes over the falls, 10^4 kg/s.

$$P = (6.0 \times 10^6\ kg/s)(9.81\ m/s^2)(53\ m) = 3.1 \times 10^9\ W = \boxed{3.1\ GW}.$$

6.71 KE in the tube of air $= \frac{1}{2}mv^2 = \frac{1}{2}\rho\,(\text{volume})\,v^2$, where ρ is the density and the

volume $= \dfrac{\pi D^2}{4}L$. The distance L = vt, so that

$KE = \dfrac{\pi}{8}\rho D^2 v^3 t$. The power is KE/t $= \boxed{\dfrac{\pi}{8}\rho D^2 v^3}$.

6.73 The increase in potential energy must come from the initial kinetic energy at the time the feet leave the ground. $mgh = \frac{1}{2}mv^2$, so $gh = \frac{1}{2}v^2$. We can find the speed v from the relation that power = $F\overline{v}$.

$P = F\overline{v}$, but from kinematics, $\overline{v} = \frac{v}{2}$, where v is the final speed and v_o is the initial speed. Then

$P = F\frac{v}{2} = ma\frac{v}{2}$. But we also know that $2as = v^2 - v_o^2$, where the initial speed v_o is 0. Thus $a = \frac{v^2}{2s}$ so that

$$P = m\frac{v^2}{2s}\frac{v}{2}.$$

This is rearranged to give $v^3 = \frac{4sP}{m}$. Use this last relationship to find v^2 and then use v^2 to find h from the first equation.

$$h = \frac{1}{2g}v^2 = \boxed{\frac{1}{2g}\left(\frac{4sP}{m}\right)^{2/3}}.$$

6.75 At 100% efficiency the work in must equal the work out. Work is given by force times distance through which it acts.
Work in = work out.

$F_{in}x_1 = F_{out}x_2$. Mechanical advantage is $\frac{F_{out}}{F_{in}} = \frac{x_1}{x_2} = \frac{50\text{ cm}}{3.0\text{ cm}} = \boxed{17}$.

6.77 (a) No friction. Work is done against the spring. Some of the force (and thus some of the work) is provided by gravity. First we need to find the extension of the spring from its unloaded equilibrium position. In equilibrium, the component of gravity down the plane is balanced by the spring force: $mg \sin\theta = kx_o$.

$$x_o = \frac{mg \sin\theta}{k} = \frac{2.0 \times 9.8 \times 0.5}{40} \text{ m} = 0.25 \text{ m}.$$

$\Delta Work_{spring} = \frac{1}{2}k\left(x^2 - x_o^2\right) = \frac{1}{2}40\text{ N/m}\left((0.35\text{ m})^2 - (0.25\text{ m})^2\right) = 1.20\text{ J}.$

$\Delta W_{grav} = mg(y - y_o) = mg(-0.10\text{ m}\sin 30) = -0.98\text{ J}.$

Net work to be supplied $= \Delta W_{spring} + \Delta W_{grav} = \boxed{+0.22\text{ J}}$.

(b) Friction. Work done against friction is $W = Fx = \mu Nx = \mu mg(\cos\theta)\,\Delta x$.

$W = 0.17(2.0\text{ kg} \times 9.80\text{ m/s}^2)(\cos 30°)(0.10\text{ m}) = 0.29\text{ J}.$

Total work to be supplied $= \Delta W_{spring} + \Delta W_{grav} + W_{friction} = \boxed{0.51\text{ J}}$.

Chapter 7

LINEAR MOMENTUM

7.1 $p = mv = (1500 \text{ kg})(115 \text{ km/h})\left(\frac{1000 \text{ m/km}}{3600 \text{ s/h}}\right) = \boxed{4.79 \times 10^4 \text{ kg·m/s}}$.

7.3 $p = mv = (1.67 \times 10^{-27} \text{ kg})(0.01)(3.0 \times 10^8 \text{ m/s})$

$p = \boxed{5.0 \times 10^{-21} \text{ kg·m/s}}$.

7.5 Use the definition of momentum, p = mv, to find the mass from knowledge of the momentum and the speed.

$m = \frac{p}{v} = \frac{350 \text{ kg·m/s}}{5.3 \text{ m/s}} = 6.6 \text{ kg}$. This is far too small a mass to be an automobile.

The answer is $\boxed{\text{no}}$.

7.7 Use the definition of momentum, p = mv, to find the speed from knowledge of the momentum and the mass. $v = \frac{p}{m}$.

(a) For an electron with mass 9.11×10^{-31} kg, the speed is

$$v = \frac{1.82 \times 10^{-26} \text{ kg·m/s}}{9.11 \times 10^{-31} \text{ kg}} = 2.00 \times 10^4 \text{ m/s} = \boxed{20.0 \text{ km/s}}.$$

(b) For an proton with mass 1.67×10^{-27} kg, the speed is

$$v = \frac{1.82 \times 10^{-26} \text{ kg·m/s}}{1.67 \times 10^{-27} \text{ kg}} = \boxed{10.9 \text{ m/s}}.$$

7.8 Momentum is defined as p = mv, and kinetic energy is defined as $KE = \frac{1}{2}mv^2$.

If we multiply and divide by m the kinetic energy term becomes

$$KE = \frac{1}{2}\frac{m^2v^2}{m} = \frac{(mv)^2}{2m} = \frac{p^2}{2m}.$$

7.9 (a) $p = mv = (109 \text{ kg})(9.86 \text{ m/s}) = \boxed{1.07 \times 10^3 \text{ kg·m/s}}$.

(b) $p = mv = (9.72 \times 10^{-3} \text{ kg})(728 \text{ m/s}) = \boxed{7.08 \text{ kg·m/s}}$.

7.11 Use Eq. (7.2) to get $F\Delta t = \Delta(mv) = m\Delta v$.

$$\Delta t = \frac{m\Delta v}{F} = \frac{0.14 \text{ kg } [32 \text{ m/s} - (-25 \text{ m/s})]}{4500 \text{ N}} = 1.8 \times 10^{-3} \text{ s} = \boxed{1.8 \text{ ms}}.$$

7.13 The impulse is $\Delta p = m\Delta v$, where $\Delta v = v - 0 = v$.

$$v = \Delta v = \frac{\Delta p}{m} = \frac{8.83 \text{ N·s}}{0.44 \text{ kg}} = \boxed{20 \text{ m/s.}}$$

7.15 The dimensions of the product of force and time are
$N \cdot s = (kg \cdot m \cdot s^{-2}) \cdot s = kg \cdot m \cdot s^{-1}$.
The dimensions of momentum are the dimensions of mass times velocity, or
$kg \cdot (m/s) = kg \cdot m \cdot s^{-1}$. They are the same.

7.17 We can calculate the change in momentum Δp from the product of the mass with the change in velocity $\Delta v = v - v_o$, where $v_o = 20$ m/s and $v = -16$ m/s (since v is 80% of v_o and is in the opposite direction to v_o). The time of contact is obtained with Eq. (7.3) as

$$\Delta t = \frac{\Delta p}{\overline{F}}, \text{ where } \overline{F} \text{ is the average force} = 8000 \text{ N}.$$

$\Delta p = m\Delta v = (1.2 \text{ kg})(-16 \text{ m/s} - 20 \text{ m/s}) = -43.2 \text{ kg}\cdot\text{m/s}$.
The force must be in the same direction as the momentum change, so take the magnitude of Δp. The time becomes

$$\Delta t - \frac{43.2 \text{ kg}\cdot\text{m/s}}{8000 \text{ N}} - 5.4 \times 10^{-3} \text{ s} = \boxed{5.4 \text{ ms}}.$$

7.19 From Example 7.2 we see that $\overline{F} = mgh/d$, so

$$\overline{F} = \frac{(0.64 \text{ kg})(9.81 \text{ m/s}^2)(0.73 \text{ m})}{0.024 \text{ m}} = \boxed{190 \text{ N}}.$$

7.21 We can use the impulse equation to find the time interval, but first we need to determine the change in velocity. If falls from some height h, but only rebounds to a height 0.60 h. Using the conservation of mechanical energy during the fall and during the bounce the kinetic energies just before and just after hitting the surface must be in the ration of the initial and final potential energies. So
$v_2{}^2 = 0.60\, v_1{}^2$. $v_2 = \sqrt{0.60}\, v_1 = 0.775\, v_1$. Note that v_1 and v_2 are in opposite directions. Using conservation of mechanical energy we get
$v_1 = \sqrt{2gh}$ where h is the initial height from which the ball is dropped.
From Eq. (7.2) we get

$$\Delta t = \frac{\Delta p}{\overline{F}} = \frac{m\Delta v}{\overline{F}} = \frac{m(v_2 - v_1)}{10 \text{ mg}} = \frac{1.775\, v_1}{10 \text{ g}} = \frac{1.775\sqrt{2gh}}{10 \text{ g}},$$

$$\Delta t = 0.1775\sqrt{\frac{2h}{g}} = 0.1775\sqrt{\frac{2 \times 1.4 \text{ m}}{9.81 \text{ m/s}^2}} = \boxed{0.095 \text{ s}}.$$

7.23 Use conservation of momentum: $p_{before} = p_{after}$
$0 = m_r v_r + m_b v_b = (3.51 \text{ kg})\, v_r + (9.72 \times 10^{-3} \text{ kg})(891 \text{ m/s})$

$$v_r = \frac{-(9.72 \times 10^{-3} \text{ kg})(891 \text{ m/s})}{3.51 \text{ kg}} = \boxed{-2.47 \text{ m/s}}.$$

7.25 Use conservation of momentum, (Eq. 7.5): $m_1v_1 + m_2v_2 = (m_1 + m_2)v$,
$(0.20 \text{ kg})(0.24 \text{ m/s}) + (0.42 \text{ kg})(0.52 \text{ m/s}) = (0.20 \text{ kg} + 0.43 \text{ kg})v$,

$$v = \frac{(0.048 + 0.22)\text{kg·m/s}}{0.62 \text{ kg}} = \boxed{0.43 \text{ m/s}}.$$

7.27 Use conservation of momentum, (Eq. 7.5): $m_1v_1 + m_2v_2 = m_1v_1' + m_2v_2'$
Here $v_1 = v_o$, $v_2 = 0$, $v_1' = v_2'$, so $m_1v_o = (m_1 + m_2)v_2'$.

The KE_{final} is $\frac{1}{2}(m_1+m_2)(v_2')^2 = \frac{1}{2}\frac{\{(m_1+m_2)(v_2')\}^2}{m_1+m_2}$.

$$\text{Final KE} = \frac{1}{2}\frac{(m_1v_o)^2}{m_1+m_2} = \frac{1}{2}m_1v_o^2\frac{m_1}{m_1+m_2}.$$

Thus $KE_{final} = \frac{m_1}{m_1+m_2}(KE_{initial})$.

$\frac{KE_{final}}{KE_{initial}} = \frac{m_1}{m_1 + m_2}$. $\boxed{\text{No}}$, the ratio of the kinetic energies depends on the masses, not on the initial speed of the incoming object.

7.29 (a) First conserve momentum with m = mass of bullet, M = mass of the pendulum, v = the initial velocity of bullet, and V = final velocity.
$mv = (m+M)V$.
Then conserve energy: $\frac{1}{2}(m+M)V^2 = (m+M)gh$, where h is the height to which the pendulum rises. Eliminate V to get

$$h = \frac{m^2v^2}{2g(m+M)^2} = \frac{(9.72 \times 10^{-3} \text{ kg})^2(728 \text{ m/s})^2}{2(9.81 \text{ m/s}^2)(1.260 \text{ g})^2} = \boxed{1.61 \text{ m.}}$$

(b) The horizontal distance can be found from the triangle below.

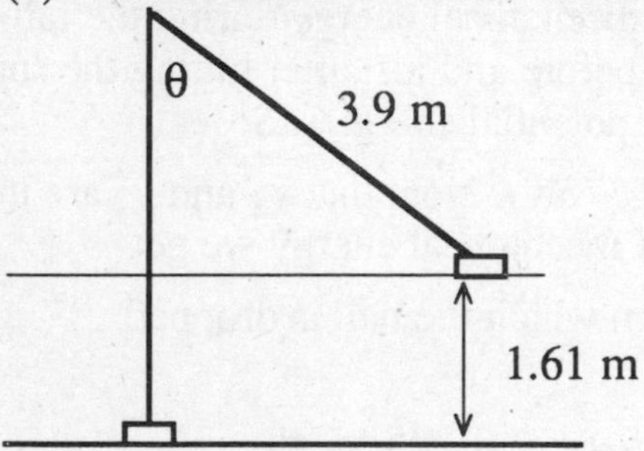

The horizontal distance is $x = 3.9 \text{ m} \sin\theta$. We can get θ from

$\cos\theta = \frac{3.9 \text{ m} - 1.61 \text{ m}}{3.9 \text{ m}}$, or $\theta = 54.0°$. So $x = 3.9 \text{ m} \sin 54.0° = \boxed{3.16 \text{ m}}$.

7.31 The velocities are oppositely directed so $v_1 = -3v_2$ and $m_1 + m_2 = 2.88$ kg.
Find m_1 and m_2. Conserve momentum: $m_1v_1 + m_2v_2 = 0$.
$(2.88 \text{ kg} - m_2)(-3v_2) + m_2v_2 = 0$.

$m_2 = \frac{3}{4}(2.88 \text{ kg}) = \boxed{2.16 \text{ kg}}$, and

$m_1 = (2.88 - 2.16) \text{ kg} = \boxed{0.72 \text{ kg}}$.

7.33 (a) Find the time for the left-hand piece to fall to hit the right-hand piece and find its velocity just before impact. Then use momentum conservation to find the velocity

of the two of them just after impact and use this velocity to find the time to reach the floor. Choose the starting position as $y = 0$. First we get
$v^2 - v_o^2 = 2gy$, where $v_o = 0$ and $y = 0.80$ m. This gives
$v = \sqrt{2(9.81 \text{ m/s}^2)0.80 \text{ m}} = 3.96$ m/s for the velocity just before impact.
The time required to fall is obtained from $v = v_o + gt$, which in this case is
$t_1 = v/g = (3.96 \text{ m/s})/(9.81 \text{ m/s}^2) = 0.404$ s.
Apply momentum conservation law to the collision (both masses are the same):
$mv = 2mv'$, so $v' = v/2 = 1.98$ m/s. Now use this velocity as the initial velocity and compute the time to fall to the bottom a distance 1.00 m away.
$v_f^2 - v'^2 = 2gy$ with $y = 1.00$ m. Insert the numbers to get
$v_f = 4.85$ m/s. The time to fall the final 1.00 m is
$t_2 = (v_f - v')/g = 0.293$ s.
The total time to fall is $t_1 + t_2 = 0.697$ s which rounds to $\boxed{0.70 \text{ s}}$.
(b) If we just consider the first piece falling the entire 1.80 m from rest without interruption the time to fall can be obtained from $y = \frac{1}{2}gt^2$, or

$$t = \sqrt{2y/g} = \sqrt{2(1.80 \text{ m})/(9.81 \text{ m/s}^2)} = 0.60 \text{ s}.$$

Thus, falling with collision takes $\boxed{\text{more time}}$ than falling without collision.

7.35 The first mass falls a distance h reaching a speed $v = \sqrt{2gh}$. (This comes from considering the conservation of mechanical energy.) Then this mass makes a perfectly inelastic collision with one of the two masses connected by the string. If we assume that the string does not stretch or break, then the effect is of the first mass m colliding with a second mass 2m. We can find the velocity immediately after collision using conservation of momentum.

$$mv = (m + 2m)v', \quad \text{so } v' = v/3 = \boxed{\frac{\sqrt{2gh}}{3}}.$$

7.37 $m_1v_1 + m_2v_2 = (m_1 + m_2)v$
Choose v_1 as the x direction and v_2 as the y direction.
x component of momentum: $m_1v_1 = (m_1 + m_2)v_x$
$v_x = (1300 \text{ kg} \times 2.25 \text{ m/s})/(3100 \text{ kg}) = 0.944$ m/s.
y component of momentum: $m_2v_2 = (m_1 + m_2)v_y$
$v_y = (1800 \text{ kg} \times 4.5 \text{ m/s})/3100 \text{ kg} = 2.61$ m/s.
$v = \sqrt{v_x^2 + v_y^2} = \boxed{2.78 \text{ m/s}}$.
v makes an angle θ with the x direction given by
$\theta = \tan^{-1}\frac{v_y}{v_x} = 70°$.

$\boxed{\theta = 70° \text{ from the direction of travel of the 1300 kg auto.}}$

7.39 In the direction parallel to train,
$mv =$ constant $= (105 \text{ kg})(7.0 \text{ m/s}) = 735$ kg·m/s.
In the direction perpendicular to the train, the sled gets momentum increase of $2m_sv_s = 2(5.0 \text{ kg})(2.0 \text{ m/s}) = 20$ kg·m/s.

Final momentum has magnitude

$p = \sqrt{p_x^2 + p_y^2} = \sqrt{735^2 + 20^2}$ kg·m/s.

The velocity magnitude is p/m = $\boxed{7 \text{ m/s}}$.

The velocity makes an angle θ with the direction of the train given by

$\theta = \tan^{-1}\dfrac{20}{735} = \boxed{1.56°}$.

7.41 In the y direction conservation of momentum gives:

$m_1v_{1y} = m_1v'_1 \sin 30° + m_2v'_2 \sin(-44°)$,

$v_{1y} = v'_1 \sin 30° + (m_2/m_1)v'_2 \sin(-44°)$,

$v_{1y} = 20 \text{ km/h} \times 0.500 + 1.2 \times 12 \text{ km/h} \times -0.695 = 0 \text{ km/h}$.

In the x direction: $m_1v_{1x} = m_1v'_1 \cos 30° + m_2v'_2 \cos(-44°)$,

$v_{1x} = v'_1 \cos 30° + (m_2/m_1)v'_2 \cos(-44°)$,

$v_{1x} = 20 \text{ km/h} \times 0.866 + 1.2 \times 12 \text{ km/h} \times 0.719 = 27.68 \text{ km/h}$.

$\boxed{v_1 = v_{1x} = 28 \text{ km/h in the x direction.}}$

7.43 $\text{Thrust} = v\dfrac{\Delta m}{\Delta t} = 2.5 \times 10^3 \text{ m/s} \times 1.3 \times 10^4 \text{ kg/s} = \boxed{3.3 \times 10^7 \text{ N}}$.

7.45 (a) The net force is the thrust (up) minus the gravitational force (down).

$$F_{net} = ma = v_r\frac{\Delta m}{\Delta t} - mg, \quad a = \frac{v_r}{m}\frac{\Delta m}{\Delta t} - g$$

$$a = \frac{2.5 \times 10^3 \text{ m/s} \times 40 \text{ kg/s}}{8.0 \times 10^3 \text{ kg}} - 9.8 \text{ m/s}^2 = \boxed{2.7 \text{ m/s}^2}.$$

(b) After 20 seconds have elapsed the mass has decreased to $m = m_o - \dfrac{\Delta m}{\Delta t}t$,

$$m = (8.0 \times 10^3) - (40 \text{ kg/s} \times 20 \text{ s}) = 7.2 \times 10^3 \text{ kg}.$$

$$a = \frac{2.5 \times 10^3 \text{ m/s} \times 40 \text{ kg/s}}{7.2 \times 10^3 \text{ kg}} - 9.8 \text{ m/s}^2 = \boxed{4.1 \text{ m/s}^2}.$$

7.47 The momenta are equal, but the masses are not. So $m_pv_p = m_Dv_D$.

$$v_D = \frac{m_p}{m_D}v_p = \frac{938.3}{1869} 2.36 \times 10^6 \text{ m/s} = \boxed{1.18 \times 10^6 \text{ m/s}}.$$

7.49 Because the cars are traveling in opposite directions, their velocities have opposite signs. Call the direction of the first car the positive direction, then $v_1 = 5.0$ km/h. The direction on the second car is negative, so $v_2 = -8.0$ km/h. The masses are equal and can be cancelled from the equation of momentum conservation, which then becomes

$$v_1 + v_2 = v_1' + v_2'.$$

After impact the cars stick together, and the final velocity is the same for each car, so $v_1' = v_2' = v'$. Thus $\quad v_1 + v_2 = 2v'$,

$$v' = \frac{v_1 + v_2}{2} = \frac{+5.0 \text{ km/h} - 8.0 \text{ km/h}}{2} = -1.5 \text{ km/h}.$$

The negative sign indicates the direction of motion.

The velocity is $\boxed{\text{1.5 km/h in the direction that the faster car was going}}$.

7.51 Conservation of momentum transverse to the initial direction of motion gives $m_1v_1\sin 30° = m_2m_2\sin 45°$.

$$\frac{m_1}{m_2} = \frac{v_1\sin 30°}{v_2\sin 45°} = \frac{4 \times 0.500}{3 \times 0.707} = \boxed{0.94}.$$

7.53 From conservation of momentum we get

$m_1v_1 = m_1v'_1 + m_2v'_2 = m_1(-v_1/20) + m_2(0.95\ v_1)$,

$$m_2 = \frac{m_1(1 + 1/20)}{0.95} = 1.1\ m_1. \quad \boxed{\frac{m_2}{m_1} = 1.1}.$$

7.55 (a) Conservation of momentum during the collision gives $mv = (m+M)V$. Conservation of mechanical energy afterwards gives $\frac{1}{2}(m+M)V^2 = (m+M)gh$, where h is the height to which the pendulum rises. Eliminate V to get

$$v = \frac{m+M}{m}\sqrt{2gh} = \frac{0.260 \text{ kg}}{0.0567 \text{ kg}}\sqrt{2(9.81 \text{ m/s}^2)(0.131 \text{ m})} = \boxed{7.35 \text{ m/s}}.$$

(b) The ball travels a horizontal distance $x = v_xt$. We can get t from motion in the vertical direction. Choosing down as positive and the 0 position as the starting point we get $y = \frac{1}{2}gt^2$ or $t = \sqrt{\frac{2y}{g}} = \sqrt{\frac{2 \times 1.12 \text{ m}}{9.81 \text{ m/s}^2}} = 0.478$ s.

The horizontal distance becomes

$x = v_xt = 7.35 \text{ m/s} \times 0.478 \text{ s} = \boxed{3.51 \text{ m}}$.

7.57 (a) The force on the rocket is the thrust (up) and gravity (down). From Newton's second law we get $ma = v_r\frac{\Delta m}{\Delta t} - mg$

$$a = \frac{(3.0 \times 10^3 \text{ m/s})(30 \text{ kg/s})}{5.0 \times 10^3 \text{ kg}} - 9.8 \text{ m/s}^2 = \boxed{8.2 \text{ m/s}^2}.$$

(b) $y = \frac{1}{2}at^2 = \frac{1}{2}(8.2 \text{ m/s}^2)(10 \text{ s})^2 = \boxed{410 \text{ m}}$.

(c) Decreased mass would mean greater acceleration and thus greater height.

Chapter 8

COMBINING CONSERVATION OF ENERGY AND MOMENTUM

8.1 At the height before it was dropped and at the maximum height of the rebound, the marble had only gravitational potential energy: $PE_i = mgh_i$, $PE_f = mgh_f$.

$$\frac{|\Delta PE|}{PE_i} = \frac{mg|(h_f - h_i)|}{mgh_i} = \frac{|1.6 - 2.0|}{2.0} = \frac{0.4}{2.0} = \boxed{0.20}.$$

The lost energy goes into sound and thermal energy.

8.3 Thrown down with a speed v_o from a height h, the ball hits the ground with a speed v given by $v^2 = v_o^2 + 2gh$.

Assume that the collision with the ground is elastic. Then the initial upward speed is also v and the ball will rise to a new height h' given by

$$mgh' = \frac{1}{2}mv^2 = \frac{1}{2}m(v_o^2 + 2gh),$$

or $h' = h + \frac{v_o^2}{2g} = 1.07 \text{ m} + \frac{(2.32 \text{ m/s})^2}{2(9.81 \text{ m/s}^2)} = 1.34 \text{ m}.$

$\boxed{\text{Yes}}$, the collision is elastic.

8.5 After three bounces, $KE = \left(\frac{1}{2}\right)^3 KE_o = 0.125\ KE_o$. The height to which it will bounce is proportional to the initial kinetic energy as rebounds from the floor. After third bounce the maximum height is

$h = 0.125\ h_o = 0.125\ (3.00 \text{ m}) = \boxed{0.375 \text{ m}}$.

8.7 Conservation of momentum: $m_1v_1 + m_2v_2 = (m_1 + m_2)V$, where $v_2 = 0$.

(a) $V = \frac{0.400 \times 0.100 \text{ kg·m/s}}{(0.400 + 0.300)\text{kg}} = \boxed{0.0571 \text{ m/s}}$.

(b) $KE(\text{lost}) = KE_i - KE_f = \frac{1}{2}m_1v_1^2 - \frac{1}{2}(m_1 + m_2)V^2,$

$KE(\text{lost}) = 2.00 \times 10^{-3}\text{ J} - 1.14 \times 10^{-3}\text{ J} = \boxed{8.59 \times 10^{-4}\text{ J}}$.

8.9 P: $mv_o - mv_o = mv_1 + mv_2 = 0$. So $v_1 = -v_2$

KE: $2\left(\frac{1}{2}mv_o^2\right) = \frac{1}{2}mv_1^2 + \frac{1}{2}mv_2^2$

So $2mv_o^2 = 2mv_1^2$, and $v_1 = \pm v_o$. Choose the – sign, then

$\boxed{v_1 = -v_o \text{ and } v_2 = +v_o}$.

8.11 For elastic collision we conserve both momentum and kinetic energy.

(momentum) $m_1v_o = m_1v_1 + m_2v_2$.

(kinetic energy) $\frac{1}{2}mv_o^2 = \frac{1}{2}m_1v_1^2 + \frac{1}{2}m_2v_2^2$.

$v_o = 4.0$ m/s, and $m_2 = 3\ m_1$.

Insert these into momentum equation to get: $4 = v_1 + 3v_2$.

From KE equation we get $16 = v_1^2 + 3v_2^2$.

Substitute for v_2: $16 = v_1^2 + 3\left(\frac{4 - v_1}{3}\right)^2$.

$4v_1^2 - 8v_1 - 32 = 0$. Solve the quadratic equation to get $v_1 = 1 \pm 3$ m/s. So $v_1 = 4$ m/s or $v_1 = -2$ m/s. The first answer corresponds to no collision. Therefore the answers are $\boxed{v_1 = -2 \text{ m/s}}$ and $\boxed{v_2 = +2 \text{ m/s}}$.

8.13 Conservation of momentum gives:
$mv_1 + mv_2 = mv_1' + mv_2' = 3.0m - 2.0m = m$. (v in units of m/s)
So, $v_1 + v_2 = 1.0$ m/s. If we write the equation for kinetic energy and divide out factors of m/2, we get
$v_1^2 + v_2^2 = (3.0)^2 + (2.0)^2 = 13.0$. Substitute for v_2 using the momentum equation:

$v_1^2 + (1.0 - v_1)^2 = 13.0$ or $2v_1^2 - 2v_1 - 12 = 0$.

$$v_1 = \frac{2 \pm \sqrt{4 + 96}}{4} = (0.5 \pm 2.5) \text{ m/s}.$$

$\boxed{v_1 = -2.0 \text{ m/s}}$ and $\boxed{v_2 = 3.0 \text{ m/s}}$.

8.15 The equation for momentum conservation is $m_1v_1 = m_1v'_1 + m_2v'_2$, where we ignore the mass of the wax. The masses of the two gliders are identical, so $m_1 = m_2$, and the mass can be factored out.
$v_1 = 4.0 = v'_1 + 3.0$; $v'_1 = 1.0$ m/s.
The energy lost is

$$KE_i - KE_f = \frac{1}{2} m \left(v_1^2 - v_1'^2 - v_2'^2\right) = \frac{1}{2} m \left(4.0^2 - 1.0^2 - 3.0^2\right).$$

$$\frac{\Delta KE}{KE_i} = \frac{16.0 - 1.0 - 9.0}{16.0} = \boxed{\frac{3}{8}}.$$

8.17 For this case, the expressions for conservation of momentum and conservation of kinetic energy are
$m_1v_1 = m_1v_1' + m_2v_2'$, and
$\frac{1}{2} m_1v_1^2 = \frac{1}{2} m_1v_1'^2 + \frac{1}{2} m_2v_2'^2$.
Solve the momentum equation for v_1' and substitute that expression into the kinetic energy equation. Solving for v_2' gives (in addition to the solution $v_2' = 0$)

$$v_2' = \frac{2 m_1}{(m_1 + m_2)} v_1.$$

The glider will go along the horizontal track a distance x until the initial kinetic energy is all dissipated as work against the frictional force F_{fr}

$$F_{fr}\, x = \mu m_2 g x = \frac{1}{2} m_2 v_2'^2.$$

Solve for x and insert the expression for v_2'.

$$x = \left(\frac{m_1}{m_1 + m_2}\right)^2 \left(\frac{2\, v_1^2}{\mu\, g}\right),$$

$$x = \left(\frac{0.283}{0.283 + 0.467}\right)^2 \left(\frac{2(0.69 \text{ m/s})^2}{(0.020)\,(9.81 \text{ m/s}^2)}\right) = 0.69 \text{ m}.$$

The answer is $\boxed{\text{no}}$, the distance the glider would go is less than the distance to the end of the track.

8.19 First, find the speed of the incident ball at height h. From conservation of mechanical energy we find

$$\frac{1}{2}(2m)v_o^2 = \frac{1}{2}(2m)v^2 + (2m)gh, \text{ which gives } v^2 = v_o^2 - 2gh.$$

Next, conserve momentum and energy for the elastic collision between the ball of mass 2m and the ball of mass m.

p: $2mv = 2mv_1 + mv_2$.

KE: $2mv^2 = 2mv_1^2 + mv_2^2$.

Factor the mass m from the equations. Then solve for v_2. First square the momentum equation to get

$(2v_1)^2 = (2v - v_2)^2$, or $4v_1^2 = 4v^2 - 4vv_2 + v_2^2$. From the energy equation:

$4v_1^2 = 4v^2 - 2v_2^2 = 4v^2 - 4vv_2 + v_2^2$, so $v_2 = \frac{4}{3}v$. The height above the floor is

found from $H = h + y$ where y is obtained from

$$mgy = \frac{1}{2}mv_2^2 = m\frac{8}{9}v^2 = m\frac{8}{9}(v_o^2 - 2gh).$$

$$H = h + \frac{8}{9g}\left(v_o^2 - 2gh\right) = \boxed{\frac{8}{9}\frac{v_o^2}{g} - \frac{7}{9}h}.$$

8.21 The protons have equal speeds. The velocity component along the initial direction is $v_x = v'_1 \cos 45°$. Conservation of momentum requires that $mv_o = 2mv_x = 2m\left(v'_1/\sqrt{2}\right)$. The final speed of each proton is $v'_1 = \boxed{v_o/\sqrt{2}}$.

8.23 Conserve momentum, noting that the masses are identical so that the angle between the emerging balls is 90° for an elastic collision.

$v_1 = v'_1 \cos 60° + v'_2 \cos 30°$

$v'_1 \sin 60° = v'_2 \sin 30°$. Solve for v'_2.

$v_1 = 0.500\, v'_1 + 0.866\, v'_2$, and $0.866\, v'_1 = 0.500\, v'_2$

$$v_1 = \frac{0.250\, v'_2}{0.866} + 0.866\, v'_2 = 1.15\, v'_2$$

So $v'_2 = 0.866\, v_1$.

$$\frac{KE_2}{KE_{1i}} = \frac{v'^2_2}{v^2_1} = (0.866)^2 = \boxed{0.75}.$$

8.25 If 1/4 of the initial energy is lost, then the final kinetic energy is 3/4 the initial kinetic energy. If the balls make equal angles with the incident direction, then they must have the same velocity and the same kinetic energy. So

$$2\left(\tfrac{1}{2}mv^2\right) = \tfrac{3}{4}\left(\tfrac{1}{2}mv_o^2\right)$$

$$v = \sqrt{\tfrac{3}{8}}\, v_o.$$

From momentum conservation we get $mv\cos\theta = \frac{1}{2}mv_o$,

$$\cos\theta = \frac{1}{2}\frac{1}{\sqrt{3/8}} - \sqrt{\tfrac{2}{3}}. \quad \boxed{\theta = 35.3°.}$$

8.27 Use Eq.(8.9) converted for units of km/h.

$$s(m) = \frac{[v(km/h)]^2}{250\mu} = \frac{(120)^2}{250 \times 0.45} = \boxed{130\text{ m}}.$$

8.29 Use Eq.(8.9) converted for units of mi/h and ft as shown in Problem 8.26.

(a) To stop on dry concrete:

$$s_1(ft) = \frac{[v(mi/h)]^2}{30\mu} = \frac{(40)^2}{30 \times 0.70} = \boxed{76\text{ ft}}.$$

(b) On wet ice the stopping distance is

$$s_2(ft) = \frac{[v(mi/h)]^2}{30\mu} = \frac{(40)^2}{30 \times 0.10} = 533\text{ ft}.$$

It takes $s_2 - s_1 = 533 - 76 = 457\text{ ft} \approx \boxed{460\text{ ft}}$ farther.

8.31 From Newton's second law, $F = \mu mg = ma$. $a = \mu g = \boxed{0.70g}$.

8.33 We call the moving car 1 and the struck car 2, giving

$m_1v_1 = (m_1 + m_2)v_1'$

Solving for v_1' and substituting into the expression for stopping a car (Eq. 3.9) gives

$$s = \left(\frac{1}{m_1 + m_2}\right)^2 \times \frac{(m_1v_1)^2}{2\mu g},$$

$$s = \left(\frac{1}{m_1 + m_2}\right)^2 \times \frac{p_1^2}{2\mu g}$$

where p is the momentum. In both cases p_1 is the same , and $(m_1 + m_2)$ is the same, so the stuck-together cars go the same distance in both cases.

8.35 Total energy = $(KE + PE)_{\text{at earth radius}} = PE_{\text{at 2 earth radii}}$,

$$E = \frac{1}{2}mv^2 - G\frac{mM_E}{R_E} = -G\frac{mM_E}{2R_E},$$

$$v = \sqrt{\frac{GM_E}{R_E}} = \sqrt{\frac{(6.67\times10^{-11})(5.98\times10^{24})}{6.37\times10^6}}\ \text{m/s}.$$

$\boxed{v = 7.91\times10^3\ \text{m/s.}}$

8.37 Total energy = $(KE + PE)_{\text{at earth radius}} = PE_{\text{at 1.5 earth radius}}$.

$$E = \frac{1}{2}mv_0^2 - G\frac{mM_E}{R_E} = -G\frac{mM_E}{\frac{3}{2}R_E},$$

$$v_0 = \sqrt{\frac{2GM_E}{R_E}\left(1-\frac{2}{3}\right)} = \boxed{6.46\times10^3\ \text{m/s}}.$$

8.39 Total energy = $(KE + PE)_{\text{at } r} = (KE + PE)_{\text{at 4 Re}}$

$$E = \frac{1}{2}mv^2 - G\frac{mM_E}{r} = \frac{1}{2}mv_0^2 - G\frac{mM_E}{4R_E},$$

$$\frac{GM_E}{r} = \frac{1}{2}\left(v_0^2 - v^2\right) - G\frac{mM_E}{4R_E},$$

$$\frac{1}{r} = \frac{v^2 - v_0^2}{2GM_E} + \frac{1}{4R_E}.$$

Insert the numerical values to get $r = 7.2\times10^6\ \text{m} \approx \boxed{1.1\ R_E}$.

8.41 Set the forces equal: $G\frac{mM_s}{R_1^2} = G\frac{mM_J}{R_2^2}$.

M_s is the mass of the sun and M_J is the mass of Jupiter. Let $R_2 = xR$ where x is the fraction of the distance R from the sun to Jupiter. Then $R_1 = (1 - x)R$.

$$\frac{M_s}{(1-x)^2R^2} = \frac{M_J}{x^2R^2},\ \text{or}\ \frac{(1-x)^2}{x^2} = \frac{M_s}{M_J}.$$

$$1 - x = \sqrt{\frac{M_s}{M_J}}\,x.\quad \text{So}\ x = \frac{1}{1+\sqrt{M_s/M_J}} = \boxed{0.03}.$$

8.43 From Eq. (8.11) we get

$$v_{esc} = \sqrt{\frac{2GM}{R}} = \sqrt{\frac{2(6.67\times10^{-11})(4.82\times10^{24})}{6.31\times10^6}},$$

$$v_{esc} = 1.01\times10^4\ \text{m/s} = \boxed{10.1\ \text{km/s}}.$$

8.45 From Eq. (8.11) we get

$$v_{esc} = \sqrt{\frac{2GM}{R}},\ \text{so}\ \frac{2GM}{R} = (v_{esc})^2.$$

Acceleration of gravity $=$ "g" $= \frac{GM}{R^2}$.

$$2\text{"g"}R = \frac{2GM}{R} = (v_{esc})^2.$$

$$R = \frac{(v_{esc})^2}{2\text{"g"}} = \frac{(3.5 \times 10^4 \text{ m/s})^2}{2 \times 11.1 \text{ m/s}^2} = \boxed{5.5 \times 10^7 \text{ m}}.$$

8.47 (a) KE + PE = constant.

$$\frac{1}{2} m v_o^2 - \frac{GmM_E}{R_E} = -\frac{GmM_E}{R_E+h}.$$

$$v_o = \sqrt{2GM_E\left(\frac{1}{R_E} - \frac{1}{R_E+h}\right)} = \sqrt{\frac{2GM_E}{R_E}\left(1 - \frac{1}{1+h/R_E}\right)}$$

For $h << R_E$ we get

$$v_o \approx \sqrt{\frac{2GM_E}{R_E}(1-(1-h/R_E))} = \sqrt{\frac{2GM_E}{R_E}\left(\frac{h}{R_E}\right)} = \boxed{\sqrt{2gh}}.$$

(b) $v_{esc} = \sqrt{2gR_E}$, so $\frac{v_o}{v_{esc}} = \sqrt{\frac{2gh}{2gR_E}} = \sqrt{\frac{h}{R_E}}$.

$$v_o = \sqrt{\frac{h}{R_E}}\, v_{esc.} = \boxed{0.088\ v_{esc}}.$$

8.49 Use the frame of reference of the spacecraft before separation. Then momentum is 0. Take the direction of the spacecraft as positive.

$(m_s + m_b)v_o = 0 = m_s v_s + m_b v_b$,

$1200\ v_s + 4800\ v_b = 0$. Thus $v_s = -4\ v_b$.

But $v_s - v_b = 100$ m/s, so $-4\ v_b - v_b = 100$ m/s or

$v_b = -20$ m/s and $v_s = +80$ m/s.

The energy released is found from the kinetic energy

$$KE = \frac{1}{2} m_s v_s^2 + \frac{1}{2} m_b v_b^2,$$

$$KE = \frac{1}{2} 1200(80)^2 \text{ J} + \frac{1}{2} 4800(-20)^2 \text{ J} = \boxed{4.8 \times 10^6 \text{ J}}.$$

8.51 Because the gliders are initially at rest, conservation of momentum gives

$0 = m_1 v_1' + m_2 v_2'$.

The stored energy of the compressed spring is converted into the kinetic energy of the gliders $\frac{1}{2} k x^2 = \frac{1}{2} m_1 v_1'^2 + \frac{1}{2} m_2 v_2'^2$.

Solving the momentum equation for v_2', substituting that expression into the kinetic energy equation, and solving for v_1' gives

$$\frac{1}{2} kx^2 = \frac{1}{2} m_1 v_1'^2 + \frac{1}{2} m_2 \frac{m_1^2}{m_2^2} v_1'^2,$$

$$v_1^2 = kx^2 \frac{m_2}{m_1}\left(\frac{1}{m_1 + m_2}\right),$$

$$v_1' = x\sqrt{k \frac{m_2}{m_1} \frac{1}{m_1 + m_2}},$$

$$v_1' = 0.0470\text{ m}\sqrt{200\text{ N/m}\frac{0.543}{0.234}\frac{1}{0.234+0.543}},$$

$v_1' = \boxed{1.15\text{ m/s}}$.

From the momentum conservation equation we get

$v_2' = -v_1'\dfrac{m_1}{m_2} = -1.15\text{ m/s}\dfrac{0.234}{0.543}\cdot\ \boxed{v_2' = -0.495\text{ m/s}}$.

8.53 (a) Work done by friction is $Fd = 3.6\times10^3\text{ N}\times0.080\text{ m} = 288\text{ J}$.

Incident KE $= \frac{1}{2}mv_o^2 = \frac{1}{2}0.036\text{ kg }(350\text{ m/s})^2 = 2.21\times10^3\text{ J}$.

KE upon emerging is $KE_{initial} - Fd = 1.92\times10^3\text{ J}$.

Emergent velocity is $v = \sqrt{2\text{ KE/m}} = \boxed{326\text{ m/s.}}$

(b) Assuming the same retarding force for each board, the number of posts is equal to the incident KE divided by the energy loss per board. Thus the number of posts penetrated is $n = \dfrac{2210\text{ J}}{288\text{ J}} = 7.67$. The answer is $\boxed{7}$. It would go through 7 but not through 8.

8.55 $KE_{initial} = PE_{spring} + W$, where W is the work done by friction.

$\frac{1}{2}mv^2 = \frac{1}{2}kx^2 + \mu mgx$.

$v^2 = \dfrac{k}{m}x^2 + 2\mu gx = 1.59\text{ m}^2/\text{s}^2$. So $v = \boxed{1.3\text{ m/s}}$.

8.57 Let m_o = mass of incident marble and m be mass of the struck marble after the chip is lost. From conservation of momentum we get

$m_o v = m_o v_1 + mv_2$, or $v_1 = \dfrac{m_o v - mv_2}{m_o}$.

From kinetic energy conservation we get

$\frac{1}{2}m_o v^2 = \frac{1}{2}m_o v_1^2 + \frac{1}{2}mv_2^2$,

$$m_o v^2 = m_o\left(\frac{m_o v - mv_2}{m_o}\right)^2 + mv_2^2,$$

$m_o v^2 = m_o v^2 - 2mvv_2 + \dfrac{m^2}{m_o}v_2^2 + mv_2^2$. Thus $m = m_o\,\dfrac{2v - v_2}{v_2}$.

$$\text{So } \Delta m = m_o - m = m_o\left[1 - \frac{2v - v_2}{v_2}\right] = 5.00\text{ g}\left[1 - \frac{99}{100}\times\frac{98}{99}\right]$$

$\Delta m = \boxed{0.100\text{ g}}$.

8.59 (a) The gravitational potential energy of a mass m due to the earth and the moon is $PE = -\dfrac{GmM_E}{r} - \dfrac{GmM_m}{R-r}$, where r is the distance from the center of the earth, and R is the distance from the center of the earth to the center of the moon. If the mass is 1-kg, the potential energy is

$$PE = -(Gm)\left(\frac{M_E}{r} + \frac{M_m}{R-r}\right),$$

$$PE = -6.67 \times 10^{-11}\left(\frac{5.98 \times 10^{24}}{r} + \frac{7.36 \times 10^{22}}{R-r}\right), \text{ where } R = 3.84 \times 10^8 \text{ m.}$$

Compute for r/R from 0.05 to 0.98. The graph is shown below.

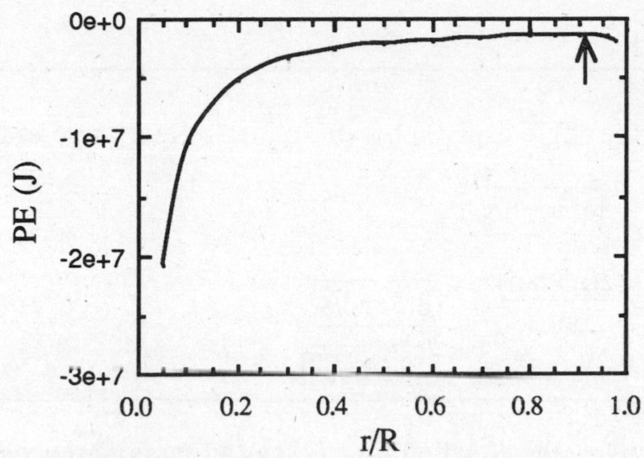

If the range of the graph is restricted we increase the detail near the maximum.

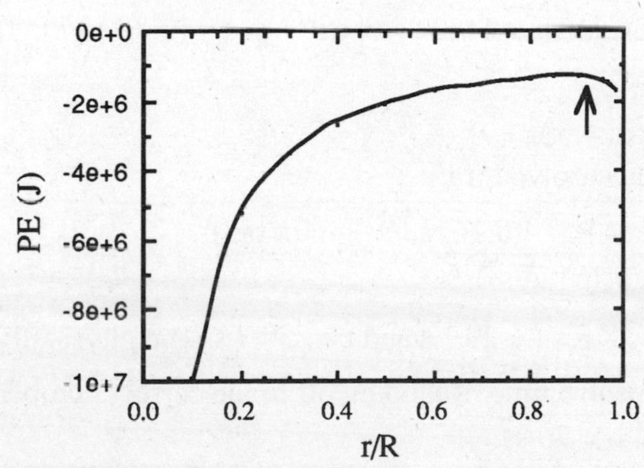

(b) The forces balance at a distance r when

$$\frac{5.98 \times 10^{24}}{r^2} = \frac{7.36 \times 10^{22}}{(R-r)^2},$$

$7.36 \times 10^{22}\, r^2 = 5.98 \times 10^{24}\,(R-r)^2.$

Take the square root of each side of the equation to get

$2.71 \times 10^{11}\, r = 2.445 \times 10^{12}\,(R-r),$

$r\,(2.71 \times 10^{11} + 2.445 \times 10^{12}) = 2.445 \times 10^{12}\, R$, so $r = \boxed{0.90\, R}$.

Note, since the roots can be either positive or negative, we can solve the equation with one side negative. In that case we get r = 1.13 R. This is also an allowed answer, but it lies beyond the moon and we were only looking for a maxima lying between the earth and the moon.

Chapter 9

RIGID BODIES AND ROTATIONAL MOTION

9.1 The angular speed is defined by $\omega = 2\pi f = \frac{2\pi}{T}$, where T is the period.

$\omega = \frac{2\pi}{13\ \text{s}} = \boxed{0.48\ \text{rad/s}}$.

9.3 (a) Starting from rest, the angle is given by $\theta = \frac{1}{2}\alpha t^2$. Rearrange to get α:

$$\alpha = \frac{2\theta}{t^2} = \frac{2(25\ \text{rad})}{(6.0\ \text{s})^2} = \boxed{1.4\ \text{rad/s}^2}.$$

(b) $\omega = \alpha t = \frac{2\theta}{t^2}t = \frac{2(25\ \text{rad})}{6.0\ \text{s}} = \boxed{8.3\ \text{rad/s}}$.

(c) $a_c = \omega^2 r = (8.3\ \text{rad/s})^2\ 0.45\ \text{m} = \boxed{31\ \text{m/s}^2}$.

9.5 This problem is solved using the equation for angle as a function of acceleration and time: $\theta = \theta_o + \omega_o t + \frac{1}{2}\alpha t^2$. Here $\omega_o = 1.6$ rad/s, $\theta = 0.32\ \text{rad/s}^2$, and $\theta - \theta_o =$ 30 revolutions = (30)(2π) radians. Then $\frac{1}{2}\alpha t^2 + \omega_o t - (\theta - \theta_o) = 0$. Insert the values for α and θ to get:

$(0.16\ \text{rad/s})t^2 + (1.6\ \text{rad/s})t - 60\pi = 0$.

Use the quadratic formula to solve for t.

$$t = \frac{-1.6\ \text{rad/s} \pm \sqrt{(1.6\ \text{rad/s})^2 - 4(0.16\ \text{rad/s}^2)(-60\pi\ \text{rad})}}{2(0.16\ \text{rad/s}^2)}$$

$t = \frac{-1.6 \pm 11.1}{0.32}$ s. This gives t = 29.7 s and t = –39.7 s. The physically sensible answer corresponds to positive time. Rounding off to the correct number of significant figures we get $\boxed{t = 30\ \text{s}}$.

9.7 First compute the total time the ball is in the air. Since $\omega = \frac{\theta}{t}$, we get

$$t = \frac{\theta}{\omega} = \frac{7.2\ \text{rev} \times 2\pi\ \text{rad/rev}}{31\ \text{rad/s}} = 1.46\ \text{s}.$$

The time to go up equals the time to come down. Consider the motion of the ball as it falls from the peak height. We can use the kinematic equation

$y - y_o = v_{oy}t + \frac{1}{2}at^2$.

At the very top its linear velocity v_{oy} is zero. The acceleration of the ball is $a = -g = -9.81\ \text{m/s}^2$. The time is one half the time found above and is 0.73 s, and $y - y_o = -h$, the height.

$-h = 0 + \frac{1}{2}(-9.81\ \text{m/s}^2)(0.73\ \text{s})^2 = \boxed{2.6\ \text{m}}$.

9.9 (a) The angular velocity is obtained from the relationship

$\omega_o = 2\pi f = 2\pi\ 800\ \text{rev/min} \times \dfrac{1\ \text{min}}{60\ \text{s}} = \boxed{83.8\ \text{rad/s}}$.

$\omega = 2\pi f = 2\pi\ 3400\ \text{rev/min} \times \dfrac{1\ \text{min}}{60\ \text{s}} = \boxed{356\ \text{rad/s}}$.

(b) The average angular acceleration is found from

$\bar{\alpha} = \dfrac{\omega - \omega_o}{t} = \dfrac{(356 - 83.8)\ \text{rad/s}}{1.20\ \text{s}} = \boxed{227\ \text{rad/s}^2}$.

(c) The total number of revolutions can be obtained from the total angle from

$\theta = \omega_o + \frac{1}{2}\alpha t^2 = 83.8\ \text{rad/s} \times 1.20\ \text{s} + \dfrac{227\ \text{rad/s}^2}{2}(1.20\ \text{s})^2$,

$\theta = 264\ \text{rad}\ \dfrac{1\ \text{rev}}{2\pi\ \text{rad}} = \boxed{42.0\ \text{rev}}$.

9.11 From the definition of torque, $\tau = rF$. So $r = \dfrac{\tau}{F} = \dfrac{35\ \text{ft·lb}}{50\ \text{lb}} = \boxed{0.70\ \text{ft}}$.

9.13 The torque applied to the handle must equal the torque exerted by the claw on the nail. $Fr = F'r'$. $100\ \text{N}(25\ \text{cm}) = F'(5\ \text{cm})$. $\boxed{F' = 500\ \text{N}}$.

9.15 First make a force diagram.

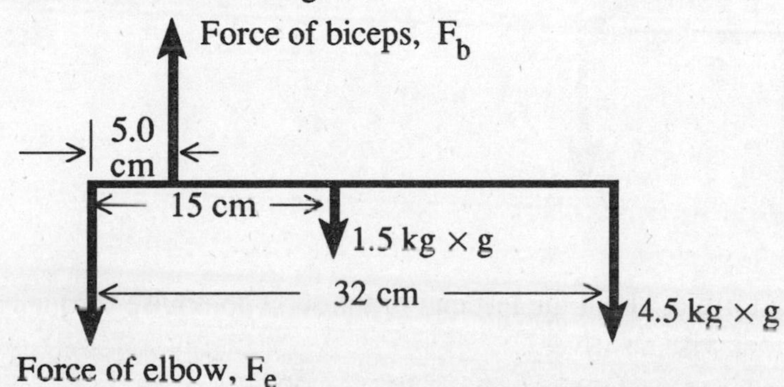

(a) Take torques about the elbow (the left end). The ccw torque is due to the muscle and the cw torques are due to the weight of the arm and the cannonball. The sum is zero.

$F_b(5.0\ \text{cm}) - (1.5\ \text{kg})(9.8\ \text{m/s}^2)(15\ \text{cm}) - (4.5\ \text{kg})(9.8\ \text{m/s}^2)(32\ \text{cm}) = 0.$

$\boxed{F_b = 330\ \text{N}}$.

(b) Take torques about the point where the biceps attaches.

$F_e(5.0\ \text{cm}) - (1.5\ \text{kg})(9.8\ \text{m/s}^2)(10\ \text{cm}) - (4.5\ \text{kg})(9.8\ \text{m/s}^2)(27\ \text{cm}). = 0.$

$\boxed{F_e = 270\ \text{N}}$.

9.17 It helps to draw the figure. The center of mass is 30 cm from the end of the bar.

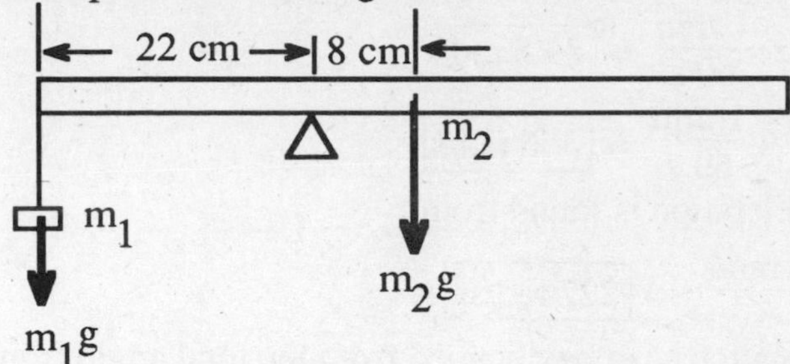

Balance the torques: $m_1gx_1 = m_2gx_2$. So $m_1x_1 = m_2x_2$.

$m_2 = \dfrac{0.75\ \text{kg} \times 22\ \text{cm}}{8\ \text{cm}} = \boxed{2.1\ \text{kg}}$.

9.19 Make a force diagram.

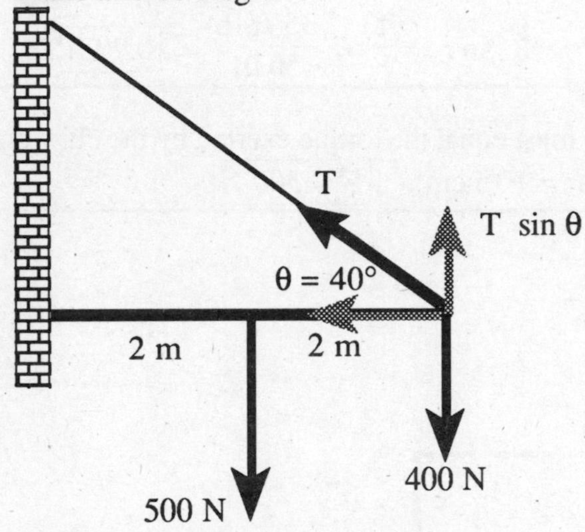

(a) $\Sigma\tau = 0$. Take torques about the left end of the rod.Clockwise torques are negative, counterclockwise positive.

$-400\ \text{N} \times 4.00\ \text{m} - 500\ \text{N} \times 2.00\ \text{m} + T \sin 40° \times 4.00\ \text{m} = 0$.

$T = \dfrac{(1600 + 1000)\ \text{N·m}}{4.00\ \text{m} \times 0.643} = \boxed{1010\ \text{N}}$.

(b) This is the same as part (a) but with $\theta = 55°$.

$T = \dfrac{2600\ \text{N·m}}{4.00\ \text{m} \times \sin 55°} = \boxed{793\ \text{N}}$.

9.21 Take torques about the feet.

$\Sigma\tau = F_{hands}\,(1.60\ \text{m}) - 75\ \text{kg}\,(9.8\ \text{m/s}^2)\ 1.00\ \text{m} = 0$.

$\boxed{F_{hands} = 460\ \text{N}}$.

Take torques about the hands.

$\Sigma\tau = -F_{feet}\,(1.60\ \text{m}) - 75\ \text{kg}\,(9.8\ \text{m/s}^2)\ 0.60\ \text{m} = 0$.

$\boxed{F_{feet} = 275\ \text{N}}$.

9.23 We can solve this problem using the definition of Young's modulus.

$Y = \dfrac{F/A}{\Delta L/L_o}$. Upon rearranging we get

$$\Delta L = \frac{F L_o}{A Y} = \frac{(5.0 \times 10^3 \text{ N})(1.3 \text{ m})}{(0.026 \text{ m})^2(9.1 \times 10^{10} \text{ N/m}^2)} = 1.1 \times 10^{-4} \text{ m} = \boxed{0.11 \text{ mm}}.$$

9.25 From the definition of the elastic modulus, $Y = \dfrac{\text{stress}}{\text{strain}}$.

If we graph stress against strain in the region where Y is constant, then the result is a straight line with a slope equal to Y. The graph would look like this:

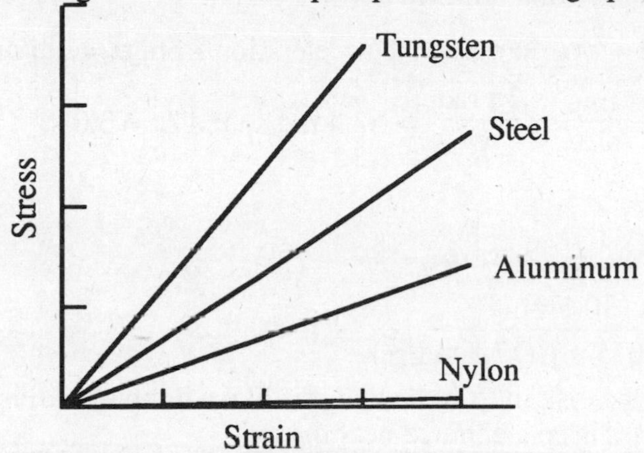

9.27 The stress is the force divided by the area. Find the cross-sectional area A, then compute the diameter.

$$\text{Breaking stress} = \frac{\text{force}}{\text{minimum area}} = \frac{F}{A}.$$

$$A = \frac{F}{\text{breaking stress}} = \frac{mg}{\text{breaking stress}} = \frac{(70.0 \text{ kg})(9.81 \text{ m/s}^2)}{11.0 \times 10^8 \text{ N/m}^2},$$

$A = 6.24 \times 10^{-7}$ m^2. But $A = \pi D^2/4$, where D is the diameter. So

$$D = \sqrt{4A/\pi} = \sqrt{4(6.24 \times 10^{-7} \text{ m}^2)/\pi} = 8.92 \times 10^{-4} \text{ m}.$$

$D = \boxed{0.892 \text{ mm}}$.

9.29 (a) For a disk the inertia I is given by $\left(\frac{1}{2}Mr^2\right)$. The torque is

$$\tau = I\alpha = \left(\tfrac{1}{2}Mr^2\right)\alpha = \tfrac{1}{2}(307 \text{ kg})(0.515 \text{ m})^2\, 1.00 \text{ rad/s}^2 = \boxed{40.7 \text{ N}\cdot\text{m}}.$$

(b) $F = \dfrac{\tau}{r} = \dfrac{40.7 \text{ N}\cdot\text{m}}{0.515 \text{ m}} = \boxed{79.1 \text{ N}}$.

(c) $F = \dfrac{\tau}{r} = \dfrac{40.7 \text{ N}\cdot\text{m}}{0.105 \text{ m}} = \boxed{388 \text{ N}}$.

9.31 The angular velocity depends on time through $\omega = \omega_o + \alpha t$. Here $\omega_o = 0$, so,

$$\omega = \alpha t = \boxed{\frac{\tau}{I}t}.$$

The tangential speed is $v = \omega R = \boxed{\frac{\tau R}{I}t}$.

9.33 We need to compute the moment of inertia of the disk. The inertia may be represented as $I = cmR^2$, where m is the mass and R the radius of the disk. The constant c is 1 for a hoop and $\frac{1}{2}$ for a uniform disk.
Since $\tau = I\alpha$, we can find $I = \tau/\alpha$. We find the acceleration from $\alpha = \Delta\omega/\Delta t$.

Here $\Delta\omega = 500 \text{ rev/min} \times \frac{1 \text{ min}}{60 \text{ s}} \times \frac{2\pi \text{ rad}}{\text{rev}} = 52.4$ rad/s, and $\Delta t = 3.0$ s.

$$\alpha = \frac{52.4 \text{ rad/s}}{3.0 \text{ s}} = 17.5 \text{ rad/s}^2.$$

$I = cmR^2$ and $I = \tau/\alpha$. So $cmR^2 = \tau/\alpha$.

$$c = \frac{\tau}{mR^2\alpha} = \frac{0.130 \text{ N·m}}{(1.37 \text{ kg})(0.075 \text{ m})^2(17.5 \text{ rad/s}^2)} = 0.97.$$

Because c is so close to 1, the disk must be so that it is like a hoop of radius R.
Disk is not uniform; it's mass is concentrated near the rim.

9.35 A normal force N produces a frictional force $F = \mu N$. This frictional force creates a torque, $\tau = rF = I\alpha$, where $\alpha = (\omega - \omega_o)/t$.

$r\mu N = I(\omega - \omega_o)/t$, where $I = \frac{1}{2}mr^2$.

$N = \frac{mr(\omega - \omega_o)}{2\mu t}$. $\omega = 0$ and $\omega_o = 2400$ rev/min $= 251$ rad/s.

$r = (19 \text{ cm})/2 = 9.5 \text{ cm} = 0.095$ m.
Also, since the wheel is slowing down the acceleration is negative which means that the force of friction is in opposition to the direction of motion. We solve for N choosing the positive (absolute) value:

$$N = \frac{(1.6 \text{ kg})(0.095 \text{ m})(251 \text{ rad/s})}{(2)(0.85)(20 \text{ s})} = \boxed{1.1 \text{ N}}.$$

9.37 Angular momentum is constant if no torques are applied.
$L = I\omega = mr^2\omega$ = constant, so $mr_f^2\omega_f = mr_i^2\omega_i$, where $\omega_f = \omega_i/2$.

$$r_f^2 = r_i^2\frac{\omega_i}{\omega_f} = 2r_i^2 = 2\,(0.80 \text{ m})^2,$$

$$r_f = \sqrt{2}\,(0.80 \text{ m}) = \boxed{1.1 \text{ m}}.$$

9.39 Use conservation of angular momentum.

$m_1\omega_1 + m_2(0) = (m_1 + m_2)\omega_f$.

$$\omega_f = \frac{m_1\omega_1}{m_1 + m_2} = \frac{1.00\text{ kg}}{1.05\text{ kg}}\left(33\tfrac{1}{3}\text{ rev/min}\right) = \boxed{31\tfrac{3}{4}\text{ rev/min}}.$$

9.41 $KE = \frac{1}{2}I\omega^2 = \frac{1}{2}\left(\frac{1}{2}mr^2\right)(2\pi f)^2$,

$KE = \frac{1}{4}(0.125\text{ kg})(6\text{ in} \times 0.0254\text{ m/in})^2\, 4\pi^2\,(45/\text{min} \times 1\text{min}/60\text{ s})^2$,

$KE = \boxed{1.6 \times 10^{-2}\text{ J}}$.

9.43 Assume the mass is concentrated at the rim. Then the moment of inertia is $I = Mr^2$.

$KE_{rot} = \frac{1}{2}I\omega^2 = \frac{1}{2}Mr^2\omega^2 = \frac{1}{2}M\,v^2$

$$KE_{rot} = \frac{1}{2}(4.0\text{ kg})\left(15\text{ km/h} \times \frac{1000\text{ m/km}}{3600\text{ s/h}}\right) = \boxed{35\text{ J}}.$$

9.45 We can solve this problem by making use of the law of conservation of mechanical energy. When the spring is wound up an amount of energy is stored as spring potential energy. When it unwinds and reaches the equilibrium position, the disk or rod is turning with a kinetic energy equal to the initial potential energy. The kinetic energy will be the same for both the rod and the disk. So,

$\frac{1}{2}I_1\omega_1^2 = \frac{1}{2}I_2\omega_2^2$, where $I_1 = \frac{1}{2}mR^2$ is the inertia of the disk and $I_2 = \frac{1}{12}mL^2$ is the inertia of the rod. The masses are the same and $L = R$.

$$\omega_2 = \omega_1\sqrt{\frac{I_1}{I_2}} = \omega_1\sqrt{\frac{12\,R^2}{2\,L^2}} = \sqrt{6}\,\omega_1\frac{R}{L}.$$

$$\omega_2 = \sqrt{6}\,\omega_1 = \sqrt{6}\;18\text{ rad/s} = \boxed{44\text{ rad/s}}.$$

9.47 Use energy conservation; the potential energy at the top equals the kinetic energy at the bottom.

For the hoop: $\frac{1}{2}mv^2 + \frac{1}{2}I\omega^2 = mgh$. $I = mr^2$ and $\omega = v/r$.

So $\frac{1}{2}mv^2 + \frac{1}{2}mv^2 = mgh$.

Hoop: $v = \sqrt{gh} = \sqrt{9.8\text{ m/s}^2 \times 2.5\text{ m}} = \boxed{4.9\text{ m/s}}$.

For the disk $I = \frac{1}{2}mr^2$ so the energy conservation equation becomes

$\frac{1}{2}mv^2 + \frac{1}{4}mv^2 = mgh$.

Disk: $v = \sqrt{\frac{4}{3}gh} = \boxed{5.7\text{ m/s}}$.

9.49 Use energy conservation; the potential energy at the top equals the kinetic energy at the bottom.

For the rolling hoop: $\frac{1}{2}mv^2 + \frac{1}{2}I\omega^2 = mgh$. $I = mr^2$ and $\omega = v/r$.

So $\frac{1}{2}mv^2 + \frac{1}{2}mv^2 = mgh$.

$v_{rolling} = \sqrt{gh}$.

For the sliding hoop: $\frac{1}{2}mv^2 = mgh$. So, $v_{sliding} = \sqrt{2gh}$.

$$\boxed{\frac{v_{rolling}}{v_{sliding}} = \frac{1}{\sqrt{2}}}.$$

9.51 Use energy conservation; the potential energy at the top equals the kinetic energy at the bottom.

For the rolling hoop: \f(\f(1,2) mv^2 + \f(1,2) $I\omega^2 = mgh$. $\omega = v/r$ and $I = mr^2$, so

$\frac{1}{2}m\omega^2r^2 + \frac{1}{2}mr^2\omega^2 = mgh$.

$$\boxed{\omega = \sqrt{\frac{gh}{r^2}}}.$$

9.53 For identical masses with one billiard ball initially at rest the momentum equation is $mv_1 = mv'_1 + mv'_2$, or

$\boxed{v_1 = v'_1 + v'_2}$.

For each ball the kinetic energy becomes

$\frac{1}{2}mv^2 + \frac{1}{2}I\omega^2 = \frac{1}{2}mv^2 + \frac{1}{2}\frac{2}{5}mv^2 = \frac{7}{10}mv^2$.

Equating initial and final energies gives $\boxed{v_1^2 = v_1'^2 + v_2'^2}$.

9.55 Assume the masses start from rest and move through a distance s. The larger mass goes down while the smaller mass goes up. If we call the initial positions zero potential energy then by energy conservation we get

$mgs - 2mgs + \frac{1}{2}mv^2 + \frac{1}{2}2mv^2 + \frac{1}{2}I\omega^2 = 0$.

$I = \frac{1}{2}mR^2$, and $\omega = v/R$. Making these substitutions and rearranging we find

$mgs = \frac{1}{2}\left(3mv^2 + \frac{I}{R^2}v^2\right) = \frac{1}{2}\left(3mv^2 + \frac{1}{2}mv^2\right) = \frac{7}{4}mv^2$.

So $v^2 = \frac{4}{7}gs$. We know from kinematics that $v^2 = 2as$. Consequently

$2as = \frac{4}{7}gs$, and so $a = \boxed{\frac{2}{7}g}$.

9.57 We can use conservation of energy to obtain an equation for the inertia of the dumbbell. Let m be the hanging mass and M be the total mass of the dumbbell.

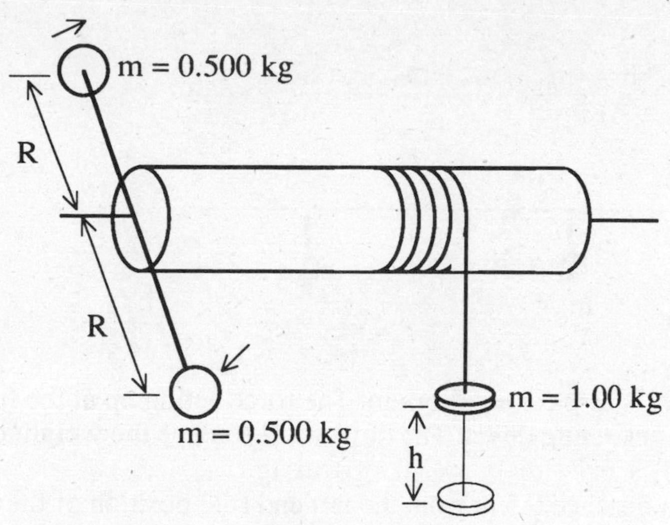

If we call the initial position the zero of potential energy then after the 1.00-kg mass falls a distance h the total energy is

$-mgh + \frac{1}{2}mv^2 + \frac{1}{2}I\omega^2 = 0.$

$I = \frac{2mgh - mv^2}{\omega^2} = MR^2$, where R is the distance from the rotation axis to one of the dumbbell masses. We are looking for d = 2R, the separation between the masses. M = 2(0.50 kg) = 1.00 kg. Also $\omega = v/r$ where r is the radius of the shaft.

$R^2 = \frac{mr^2}{M}\left(\frac{2gh - v^2}{v^2}\right) = \frac{mr^2}{M}\left(\frac{2gh}{v^2} - 1\right)$

$d = 2R = 2r\sqrt{\frac{m}{M}\left(\frac{2gh}{v^2} - 1\right)}$,

$d = 2(0.060 \text{ m})\sqrt{(1.00)\left(\frac{2(9.81 \text{ m/s}^2)(1.00 \text{ m})}{(2.68 \text{ m/s})^2} - 1\right)}$

d = 0.158 m = $\boxed{15.8 \text{ cm}}$.

9.59 To lift the 20-kg mass at constant speed, we need a force equal and opposite to the force of gravity. This is a force

$F = mg = 20 \text{ kg} \times 9.80 \text{ m/s}^2 = 196 \text{ N}.$

Equation (9.9) applies here with $\theta = 90°$, so

$\tau = r F \sin 90° = rF,$

$\tau = 0.050 \text{ m} \times 196 \text{ N} = \boxed{9.8 \text{ N}\cdot\text{m}}$.

9.61 Start with a diagram.

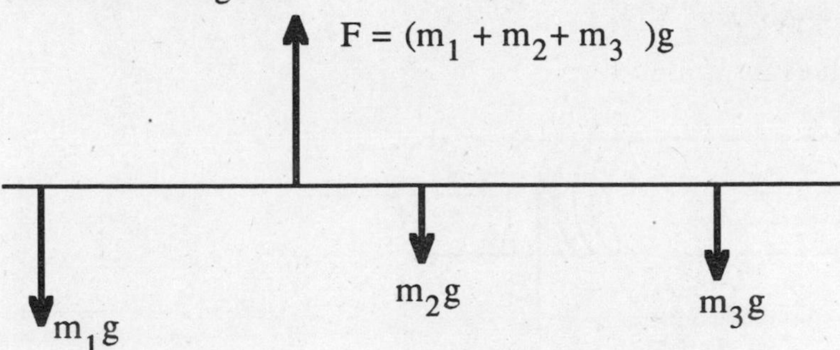

The sum of the torques is zero for equilibrium. The force acting up at the fulcrum is just the sum of the forces acting down. The downward force is the weight of the three masses. $F = (m_1 + m_2 + m_3)g = (7.0\text{ kg} + m_3)g$.
Taking torques from a distance 0.5 m from the left end (the position of the force m_2g) we get
$\Sigma\tau = (5.0\text{ kg})(g)(0.45\text{ m}) - [(7.0\text{ kg} + m_3)(g)](0.15\text{ m}) - (m_3)(g)(0.35\text{ m}) = 0.$
We can divide by g to get
$2.25\text{ kg·m} - 1.05\text{ kg·m} - (0.15\text{ m} + 0.35\text{ m})\, m_3 = 0,$
$\boxed{m_3 = 2.4\text{ kg}}$.

9.63 The sum of the torques is zero for equilibrium.
$\Sigma\tau = 0 = x_1(400\text{ N}) - (x_1 + x_2)\, F.$
$F = 400\text{ N}\dfrac{x_1}{4x_1} = \boxed{100\text{ N}}$.

9.65 When torques about the point of contact equal zero, the yo-yo slides without rotating. From the diagram the torque is zero when $\sin\theta = r/R$.

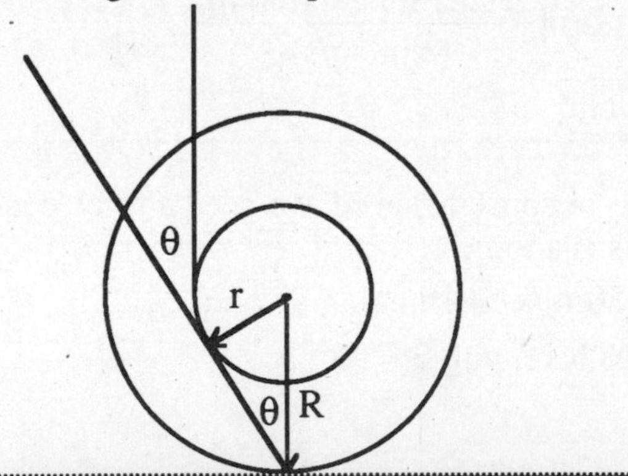

If the string is pulled at an angle $\theta = \sin^{-1}(r/R)$, the force is directed through the point of contact of the yo-yo and the table, so the net torques are zero.

9.67 Refer to the figure. A free body diagram of the mass m leads to the equation
$mg - T_1 = ma.$
A free body diagram for the mass 2m leads to
$2mg - T_2 = 2ma.$
A free body diagram for the cylinder yields the equation for torques

$(T_1 + T_2)R = I\alpha = \left(\frac{1}{2}MR^2\right)\left(\frac{a}{R}\right)$

Combine these three equations to get

$3mg - (T_1 + T_2) = 3mg - \frac{1}{2}Ma = 3ma.$

$\left(3m + \frac{1}{2}M\right)a = 3mg,$

$\boxed{a = \frac{6m}{6m + M}g}.$

9.69 Let I_1 and I_2 be the moments of inertia of the solid cylinder and the hollow cylinder, respectively. Then $I = I_1 = \frac{1}{2}mr^2$ and $I_2 = mr^2 = 2I$. The initial kinetic energy is

$KE_i = \frac{1}{2}mv^2 + \frac{1}{2}I\omega^2 = \frac{3}{4}mv^2$. The final kinetic energy is

$KE_f = \frac{3}{4}mv_1'^2 + mv_2'^2$. Conservation of energy gives

$\frac{3}{4}mv^2 = \frac{3}{4}mv_1'^2 + mv_2'^2$ or $v^2 = v_1'^2 + \frac{4}{3}v_2'^2.$

Conservation of momentum gives

$mv = mv'_1 + mv'_2$, or $v = v'_1 + v'_2$. Solve for v'_1 and insert into the energy equation to find v'_2.

$v^2 = (v - v'_2)^2 + \frac{4}{3}v_2'^2 = v^2 - 2vv'_2 + v_2'^2 + \frac{4}{3}v_2'^2.$

$2v = \frac{7}{3}v'_2$, so $v'_2 = \frac{6}{7}v = \frac{6}{7}0.25 \text{ m/s} = \boxed{0.21 \text{ m/s}}$.

9.71 The potential energy it has when released will be converted to kinetic energy just before striking the floor.

$mgh = \frac{1}{2}mv^2 + \frac{1}{2}I\omega^2 = \frac{1}{2}mv^2 + \frac{1}{2}\frac{1}{2}Mr^2(v/r)^2,$

$mgh = \frac{1}{2}mv^2 + \frac{1}{4}Mv^2.$

$v = \sqrt{\frac{mgh}{0.5m + 0.25M}} = \sqrt{\frac{(0.70\text{kg})(9.8 \text{ m/s}^2)(1.2 \text{ m})}{(0.70/2 + 1.3/4) \text{ kg}}} = \boxed{3.5 \text{ m/s}}.$

Chapter 10

FLUIDS

10.1 (a) The volume of the wood is $V = 0.10\ m \times 0.30\ m \times 0.055\ m = 1.65 \times 10^{-3}\ m^3$.

The density is $\rho = \frac{m}{V} = \frac{0.320\ kg}{1.65 \times 10^{-3}\ m^3} = \boxed{0.19 \times 10^3\ kg/m^3}$.

(b) From comparison with Table 10.2, the wood is $\boxed{\text{balsa}}$.

10.3 We can use the relation that $14.7\ lb/in^2 = 101.3\ kPa$. Then forming ratios we get

$$\frac{35\ lb/in^2}{14.7\ lb/in^2} = \frac{P}{101.3\ kPa}, \quad P = \boxed{240\ kPa}.$$

10.5 $P = \rho gh = (1000\ kg/m^3)(9.8\ m/s^2)(726\ ft)/(3.28\ ft/m) = \boxed{2170\ kPa}$.

10.7 $P = 1000\ atm = 1000 \times 101\ kPa$ and

$P = \rho gh = (1.026 \times 10^3\ kg/m^3)(9.81\ m/s^2)h$. So

$$h = \frac{1.01 \times 10^8\ N/m^2}{(1.026 \times 10^3\ kg/m^3)(9.81\ m/s^2)} = \boxed{10.1\ km}.$$

10.9 (a) $P = \frac{F}{A} = \frac{0.5 \times 530\ N}{6.5\ cm^2 \times 10^{-4}\ m^2/cm^2} = \boxed{410\ kPa}$.

For $A = 1.0\ cm^2$, $P = \frac{0.5 \times 530\ N}{10^{-4}\ m^2/cm^2} = \boxed{2650\ kPa}$.

(b) Pressure under elephant foot is $P = \frac{\frac{1}{4}37000\ N}{\pi(0.19\ m)^2} = \boxed{82\ kPa}$.

The pressure under the woman's heel is equal to the pressure of thirty two elephants stacked on top of each other.

10.11 $P = \rho_1 gh_1 + \rho_2 gh_2 = (\rho_1 + \rho_2)(9.81\ m/s^2)(0.500\ m)$, since $h_1 = h_2 = 0.500\ m$.

$P = (1.26 + 1.00)(10^3\ kg/m^3)(9.81\ m/s^2)(0.500\ m) = \boxed{11.1\ kPa}$.

10.13 The force is the product of pressure and area. From Pascal's law we get

$$F_2 = PA_2 = \frac{F_1}{A_1}A_2 = 180\ N\left(\frac{5.4\ cm}{1.0\ cm}\right)^2 = \boxed{5200\ N}.$$

10.15 From Pascal's law, the pressure is transmitted undiminished, so that the force on the piston lifting the car is given by

$F = PA = 700\ kPa \times 0.280\ m^2 \times 1000\ Pa/kPa = \boxed{1.96 \times 10^5\ N}$.

10.17 Refer to Fig. 10.30. Since mercury is so heavy, calculate the height of a column of mercury that generates a pressure equal to the 20-cm column of water

$P = \rho gh = (998 \text{ kg/m}^3)(g)(0.20 \text{ m}) = (1.36 \times 10^4 \text{ kg/m}^3)(g)(y)$

$y = \frac{998}{1360} 0.20 \text{ m} = 0.015 \text{ m} = 1.5 \text{ cm}.$

The difference in heights of the two surfaces is $\Delta y = (20 - 1.5) \text{ cm} = \boxed{18.5 \text{ cm}}$.

10.19 The buoyant force of the water just equals the weight of the cube: $F_b = mg = \rho Vg$. The buoyant force is equal to the weight of the volume of water displaced. Since 30% of the cube is above water, 70% of the volume is below the water surface. $F_b = \rho_{water}(0.70 \text{ V})g$. Combining these two equations for F_b gives $\rho = \boxed{0.70 \text{ g/cm}^3}$.

10.21 The effective weight w_{eff} is the weight less the buoyant force:

$w_{eff} = mg - \text{buoyant force}$

$w_{eff} = mg - \rho_o Vg$. The volume V is found from $\rho = m/V$, so

$w_{eff} = mg - \rho_o(m/\rho)g = mg(1 - \rho_o/\rho)$.

10.23 Density of hot air is $\rho_{hot} = 1.29\left(\frac{273}{317}\right) \text{kg/m}^3 = 1.111 \text{ kg/m}^3$.

Density of air at 21° is $\rho = 1.29\left(\frac{273}{294}\right) \text{kg/m}^3 = 1.198 \text{ kg/m}^3$.

$$V = \frac{m}{\rho - \rho_{hot}} = \frac{50 \text{ kg}}{(1.198 - 1.111)\text{kg/m}^3} = \boxed{575 \text{ m}^3}.$$

We kept more significant figures in the intermediate steps and waited until computing the final answer to round off in order to minimize round off error.

10.25 (a) The reading on the spring scale B is due to the weight of the aluminum block minus the buoyant force. If the scale is calibrated to read in kg then we divide the force by g. Thus the scale reading is

$\text{scale B} = V_{block}\rho_{Al} - \frac{1}{2} V_{block}\rho_{water} = 10^3 \text{ cm}^3 (2.70 - \frac{1}{2} 1.00) \text{ g/cm}^3 = \boxed{2.2 \text{ kg}}$.

(b) Reading on scale A is due to the sum of the weight of the container of water plus the buoyant force on the aluminum cube.

$\text{scale A} = 1.0 \text{ kg} + \frac{1}{2} V_{block}\, \rho_{water} = 1.0 \text{ kg} + \frac{1}{2} 10^3 \text{ cm}^3\ 1.00 \text{ g/cm}^3 = \boxed{1.5 \text{ kg}}$.

10.27 (a) Find the volume from the buoyant force, where the buoyant force is the difference between the weight in air and the weight in water.

$F_b = (0.475 \text{ kg} - 0.437 \text{ kg})g = \rho_{water} V g$.

$V = \frac{0.038 \text{ kg}}{1.00 \times 10^3 \text{ kg/m}^3} = 3.8 \times 10^{-5} \text{ m}^3$. Now compute the density.

$\text{Density of crown} = \frac{m}{V} = \frac{0.475 \text{ kg}}{3.8 \times 10^{-5} \text{ m}^3} = 12.5 \times 10^3 \text{ kg/m}^3$.

Density of gold is $19.3 \times 10^3 \text{ kg/m}^3$.

The crown is not solid gold.

(b) crown mass = $\rho_{average}V = \rho_{gold}V_{gold} + \rho_{lead}V_{lead}$.
Assume that $V_{gold} + V_{lead} = V$. Then
$\rho_{gold}V_{gold} + \rho_{lead}(V - V_{gold}) = \rho_{average}V$,
$V_{gold}(\rho_{gold} - \rho_{lead}) = (\rho_{average} - \rho_{lead})V$.

$$\frac{m_{gold}}{m} = \left(\frac{\rho_{gold}}{\rho_{average}}\right)\left(\frac{V_{gold}}{V}\right) = \left(\frac{\rho_{gold}}{\rho_{average}}\right)\left(\frac{\rho_{average} - \rho_{lead}}{\rho_{gold} - \rho_{lead}}\right)$$

$$\frac{m_{gold}}{m} = \left(\frac{19.3}{12.5}\right)\left(\frac{12.5 - 11.4}{19.3 - 11.4}\right) = \boxed{21.5\%}.$$

10.29 (a) We can calculate the vol;ume of thewater from the magnitude of the buoyant force. The buoyant force = $F_b = (9.34 \text{ kg} - 8.84 \text{ kg})g = 0.50 \text{ kg} \times g = \rho_{water}Vg$.

$$V = \frac{F_b}{\rho_{water}g} = \frac{0.50 \text{ kg}}{\rho_{water}} = \frac{0.50 \text{ kg}}{1.00 \times 10^3 \text{ kg/m}^3} = \boxed{0.50 \times 10^{-3} \text{ m}^3.}$$

(b) $$\rho_{uranium} = \frac{m}{V} = \frac{9.34 \text{ kg}}{0.5 \times 10^{-3} \text{ m}^3} = \boxed{18.7 \times 10^3 \text{ kg/m}^3}$$

(c) $$\rho_{liquid} = \frac{F_b}{Vg} = \frac{(9.34 - 2.54)\text{kg}}{0.50 \times 10^{-3} \text{ m}^3} = \boxed{13.6 \times 10^3 \text{ kg/m}^3}.$$

(d) The liquid is $\boxed{\text{mercury}}$.

10.31 From the definition of surface tension we get $\gamma = \frac{F}{2C} = \frac{F}{2\pi d}$.

$$F = \gamma 2\pi d = (72 \times 10^{-3} \text{ N/m})(2\pi)(0.080 \text{ m}) = \boxed{3.6 \times 10^{-2} \text{ N}}.$$

10.33 From Example 10.6, the height is given by

$$h = \frac{2\gamma \cos\theta}{r\rho g} = \frac{2(72 \times 10^{-3} \text{ N/m})(1)}{(5 \times 10^{-6} \text{ m})(10^3 \text{ kg/m}^3)(9.8 \text{ m/s}^2)} = \boxed{2.9 \text{ m}}.$$

10.35 The surface tension can be thought of as the energy per unit area of surface. So, the surface energy = $\gamma A = \gamma\pi D^2$. If the diameter increases by a factor of two, then the surface energy increases by a factor of $\boxed{4}$.

10.37 Assume constant height. Let subscript 1 refer to the 5.0-cm pipe and 2 refer to the 2.0-cm pipe. Then find the speed of the water using equation of continuity.

$v_1 = \frac{A_2}{A_1} v_2 = \left(\frac{2.0}{5.0}\right)^2 3.0 \text{ m/s} = 0.48 \text{ m/s}$. Apply Bernoulli's equation to get

$$\Delta p = p_2 - p_1 = \tfrac{1}{2}\rho\left(v_1^2 - v_2^2\right) = 500\left((0.48)^2 - (3.0)^2\right) \text{ Pa},$$

$$p_2 - p_1 = \boxed{-4.4 \text{ kPa}}.$$

10.39 Find the velocity using the equation of continuity. $A_1v_1 = A_2v_2$. So $v_2 = \frac{A_1}{A_2}v_1$

$v_2 = \frac{1.0 \times 10^{-2}\ m^2}{4.0 \times 10^{-4}\ m^2} 0.50\ m/s = \boxed{12.5\ m/s}$. Now apply bernoulli's equation.

$p_2 = p_1 + \frac{1}{2}\rho\left(v_1^2 - v_2^2\right) = \boxed{4.2 \times 10^5\ Pa}$.

10.41 Let the water surface be a height H above the ground and let the hole be a distance h below the water surface. The speed of the water emerging from the hole is $v = \sqrt{2gh}$. The time to reach the ground is obtained from $\frac{1}{2}gt^2 = H - h$, or $t = \sqrt{2(H-h)/g}$. The horizontal distance travelled is

$x = vt = \left(\sqrt{2gh}\right)\left(\sqrt{2(H-h)/g}\right) = \sqrt{(2gh)2(H-h)/g} = 2\sqrt{h(H-h)}$.

If $h = H/2$, then $x = h$. If $h = 0$, $x = 0$. If $h = H$, $x = 0$.

If we square x we get $x^2 = 4h(H-h)$. When $h = H/2$ we get $x^2 = 4(H/2)(H/2) = H^2$.

Suppose $x = \frac{H}{2} + \Delta$.

Then, $x^2 = 4\left(\frac{H}{2} + \Delta\right)\left(H - \frac{H}{2} + \Delta\right) = 4\left(\frac{H}{2} + \Delta\right)\left(\frac{H}{2} - \Delta\right) = H^2 - 4\Delta^2$.

Note that this result is less than H^2 for both positive and negative Δ. Thus the maximum range is for $h = H/2$.

10.43 From Poiseuille's law, $p_1 - p_2 = \frac{8Q\eta L}{\pi r^4}$.

If the pressure drop is constant, then $\frac{8Q_1\eta L}{\pi r_1^4} = \frac{8Q_2\eta L}{\pi r_2^4}$.

So, $\frac{Q_2}{Q_1} = \frac{r_2^4}{r_1^4} = (0.95)^4 = 0.81$. Flow is reduced by $\boxed{19\%}$.

10.45 From Poiseuille's law, $P_1 - P_2 = \frac{8Q\eta L}{\pi r^4}$.

$Q = \frac{\Delta P\ \pi r^4}{8\eta L} = \frac{(1.0 \times 10^4\ Pa)(\pi)(0.020\ m)^4}{8(1.55\ Pa{\cdot}s)(0.50\ m)} = \boxed{8.1 \times 10^{-4}\ m^3/s}$.

10.47 Use Eq. (10.12) derived from Stokes's law.

$v_t = \frac{2r^2g}{9\eta}(\rho - \rho')$. Here $\eta = 1.49$ Pa·s, $r = 4.0 \times 10^{-3}$ m and

$\rho - \rho' = (7.8 - 1.26) \times 10^3$ kg/m^3.

$v_t = \boxed{0.15 \text{ m/s}}$.

10.49 From Eq. (10.12) we get $\eta = \dfrac{2r^2g(\rho - \rho')}{9v_t}$

(a) $\eta = \dfrac{2(0.00785 \text{ m})^2 9.8 \text{ m/s}^2 (2.5 - 1.2)\times 10^3 \text{ kg/m}^3}{9(0.121 \text{ m}/45 \text{ s})}$

$\eta = \boxed{65 \text{ Pa·s}}$.

(b) $\eta = \boxed{7.2 \text{ Pa·s}}$.

10.51 The drag force is equal to the gravitational force. If the v^2 term dominates we get,

$F = mg = cv_t^2$. So $v_t = \sqrt{\dfrac{mg}{c}}$.

$$v_t = \sqrt{\frac{4.0 \times 10^{-4} \text{ kg} \times 9.8 \text{ m/s}^2}{9 \times 10^{-3} \text{ kg/m}}} = \boxed{0.66 \text{ m/s}}.$$

10.53 If the v^2 term dominates we get, $F = cv^2$. So $F_2 = F_1 \left(\dfrac{v_2}{v_1}\right)^2$.

$F_1 = 1.00 \times 10^5$ N at $v_1 = 750$ km/h.

At 800 km/h, $F_2 = 1.00 \times 10^5 \text{ N} \left(\dfrac{800}{750}\right)^2 = \boxed{1.14 \times 10^5 \text{ N}}$.

At 600 km/h, $F_2 = 1.00 \times 10^5 \text{ N} \left(\dfrac{600}{750}\right)^2 = \boxed{6.40 \times 10^4 \text{ N}}$.

$\boxed{\text{Fuel consumption is proportional to } v^2.}$

10.55 The terminal speed v_t is proportional to g according to Stokes's law.

Convert 3.0 grams/hour to grams per day: 3.0 g/h = 3.0 × 24 g/day = 72 g/day.

To achieve this rate we need an acceleration of 72 g.

$a_c = 72g = r\omega^2 = r(2\pi f)^2$.

$$f = \frac{1}{2\pi}\sqrt{\frac{72 \times 9.8 \text{ m/s}^2}{0.05 \text{ m}}} = \boxed{19 \text{ Hz}}.$$

10.57 F_d = drag force = mg at terminal speed. We can express the drag force in terms of Newton's second law.

$F_d = k\dfrac{\Delta p}{\Delta t} = k\dfrac{\Delta(mv)}{\Delta t} = k\dfrac{\Delta m}{\Delta t}v$. Here Δm represents the mass of air displaced in time Δt. If the object falls a distance h in a time t, then

$\frac{\Delta m}{\Delta t} = \frac{\rho A h}{t} = \rho A v$. Now equating gravitational force to drag force gives

$mg = k\rho A v^2$. Upon rearranging we find $\boxed{v = \sqrt{\frac{mg}{kA\rho}}}$.

10.59 $[Re] = \frac{[\rho][v][D]}{[\eta]} = \frac{(kg/m^3)(m/s)(m)}{Pa\cdot s} = \frac{kg/(m\cdot s)}{Pa\cdot s}$.

$Pa\cdot s = \frac{N}{m^2}\cdot s = \frac{kg\cdot m}{m^2\cdot s^2}\cdot s = \frac{kg}{m\cdot s}$.

So $[Re] = \frac{kg/(m\cdot s)}{kg/(m\cdot s)} = 1$.

10.61 $Re = \frac{\rho v D}{\eta} \approx 2000$ for laminar flow.

$\frac{\rho v_1 D_1}{\eta} = \frac{\rho v_2 D_2}{\eta}$, so $\frac{v_1}{v_2} = \frac{D_2}{D_1}$.

The flow rate is Av, which is proportional to D^2v. Thus

$D_2^2 v_2 = 2D_1^2 v_1$.

$\frac{v_1}{v_2} = \frac{D_2^2}{2D_1^2} = \frac{D_2}{D_1}$. So $D_2 = 2D_1 =$ $\boxed{\text{4-cm diameter}}$.

10.63 From Eq. 10.15 we find that $Re = \frac{\rho v D}{\eta} \leq 2000$, for laminar flow.

$\eta = 1 \times 10^{-3}$; $D = 0.75$ in. $= 0.019$ m; $\rho = 998$ km/m^3.

$v = \frac{(2000)(1.0 \times 10^{-3}\ Pa\cdot s)}{(998\ kg/m^3)(0.019\ m)} = 0.105$ m/s.

Flow rate $= vA = \frac{Re\,\eta}{\rho D}\frac{\pi D^2}{4} = \frac{Re\,\eta\pi D}{4\rho}$

Flow rate $= \frac{(2000)(1.0 \times 10^{-3}\ Pa\cdot s)\pi(0.019\ m)}{4(998\ kg/m^3)}$

Flow rate $= 3.0 \times 10^{-5}\ m^3/s \times 3600\ s/h =$ $\boxed{0.11\ m^3/h}$.

10.65 $P = \rho g h = (1.026 \times 10^3\ kg/m^3)(9.81\ m/s^2)\frac{4000\ ft}{3.28\ ft/m} =$ $\boxed{1.23 \times 10^7\ Pa}$.

$P = \rho g h = (1.026 \times 10^3\ kg/m^3)(9.81\ m/s^2)\frac{35000\ ft}{3.28\ ft/m} =$ $\boxed{1.07 \times 10^8\ Pa}$.

10.67 First find the area using the equation of continuity: $A_1v_1 = A_2v_2$.

So $A_2 = \frac{v_1}{v_2} A_1 = \frac{2.0}{3.0} 1.0 \times 10^{-2}\ m^2$, $A_2 = \boxed{6.7 \times 10^{-3}\ m^2}$.

Now use Bernoulli's equation to find the pressure. Note the heights are the same.

$$p_2 = p_1 + \frac{1}{2}\rho\left(v_1^2 - v_2^2\right) = \left[1.5 \times 10^4 + 500(4.0 - 9.0)\right]\ Pa$$

$p_2 = \boxed{1.25 \times 10^4\ Pa}$.

10.69 $P = \rho gh = (1030\ kg/m^3)(9.8\ m/s^2)(4000\ m) = 4.04 \times 10^7\ Pa$.

$A = \pi(D/2)^2 = 7.31 \times 10^{-2}\ m^2$.

$F = PA = \boxed{2.95 \times 10^6\ N}$.

10.71 Flow rate $= Av = \pi \frac{D^2}{4} v = 30\ L/min = 0.50\ L/s$.

$1\ L = 10^3\ cm^3 = 10^{-3}\ m^3$.

(a) $v = \dfrac{4 \times 0.50 \times 10^{-3}\ m^3/s}{\pi(0.004\ m)^2} = \boxed{40\ m/s}$.

(b) $x = vt$ and $y = \frac{1}{2}gt^2 = 1.1\ m$.

$t = \sqrt{2(1.1\ m)/(9.8\ m/s^2)} = 0.47\ s$.

$x = (40\ m/s)(0.47\ s) = \boxed{19\ m}$.

10.73 The speed can be determined with the equation given in the text.

$$v = b\left[\frac{2(\rho' - \rho)gh}{\rho(a^2 - b^2)}\right]^{1/2}$$

$$v = (0.025\ m)^2\left[\frac{2(1.36 \times 10^4 - 998)\ (9.8\ m/s^2)(0.05\ m)}{998\ (0.05^4 - 0.025^4)m^4}\right]^{1/2}$$

$v = \boxed{0.91\ m/s}$.

10.75 The weight must be supported by the buoyant force.

$mg = (\rho - \rho')Vg$.

Assume outside air temperature of 20°C, which has density

$$\rho = \rho_o \frac{273}{T + 273} = 1.29\ kg/m^3 \frac{273}{20 + 273} = 1.202\ kg/m^3.$$

$$\rho' = \rho - m/V = 1.20\ kg/m^3 - \frac{475\ kg}{2180\ m^3} = 0.984\ kg/m^3.$$

$\rho' = \rho_o \dfrac{273}{T + 273}$. Now solve for T.

$$T = 273\left(\frac{\rho_o}{\rho'} - 1\right) = \boxed{84.9°C}.$$

10.77 (a) The triangular prism has a volume given by the area of the triangular face times the thickness of the prism. $V = \frac{1}{2}bht$, where t is the thickness, h the height, and b the width of the base.

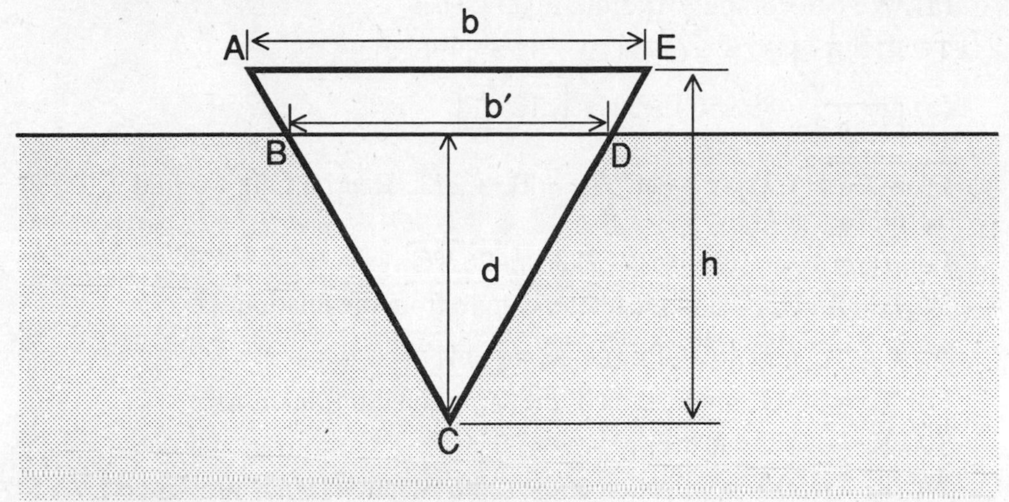

The submerged volume is given by $V_s = \frac{1}{2}b'dt$.

The weight of the ice prism is $\rho_i Vg$ and the buoyant force is the weight of the displaced water, $\rho_w V_s g$. When the prism floats these two forces are equal in magnitude so

$\rho_i Vg = \rho_w V_s g$. Upon rearanging and eliminating the common factor of g we get

$V_s/V = \boxed{\rho_i/\rho_w}$.

(b) Substituting for the volumes we find

$\rho_i \frac{1}{2} bhtg = \rho_w \frac{1}{2} b'dtg$. We can eliminate the common factors to get

$\rho_i bh = \rho_w b'd$, or

$$\frac{b'd}{bh} = \frac{\rho_i}{\rho_w}.$$

Notice that triangle ACE is similar to triangle BCD so that the ratio of base to height is the same for each triangle. Thus $b'/b = d/h$. Inserting this into the equation above leads to

$\frac{d^2}{h^2} = \frac{\rho_i}{\rho_w}$. Taking square roots we get

$d/h = \boxed{\sqrt{\rho_i/\rho_w}}$.

Chapter 11

THERMAL PHYSICS

11.1 (a) We can compute T_F from Eq.(11.1).

$T_F = \frac{9}{5}T_C + 32 = \frac{9}{5}(57.8°C) + 32 = \boxed{136°F}$.

(b) $T_F = \frac{9}{5}(-89.2°C) + 32 = \boxed{-129°F}$.

11.3 From Eq.(11.2) we have: $T_K = T_C + 273$. Rearranging we find $T_C = T_K - 273$.

For $T_K = 10$ K; $T_C = 10 - 273 = \boxed{-263°C}$.

We can find the Fahrenheit temperature from the relation

$T_F = \frac{9}{5}T_C + 32 = \frac{9}{5}(-263) + 32 = \boxed{-441°F}$.

For $T_K = 300$ K application of the same equations leads to

300 K = $\boxed{27°C = 81°F}$.

For $T_K = 450$ K application of the same equations leads to

450 K = $\boxed{177°C = 351°F}$.

11.5 Using Eq.(11.2) we get $T_K = T_C + 273.15 = -195.8 + 273.15 = \boxed{77.35 \text{ K}}$.

Then using Eq.(11.1) we find

$T_F = \frac{9}{5}T_C + 32 = \frac{9}{5}(-195.8°C) + 32 = \boxed{-320.4°F}$.

11.7 Start with Eq.(11.1): $T_F = \frac{9}{5}T_C + 32$. Rearrange Eq.(11.2) to get $T_C = T_K - 273.15$. Now insert this expression for T_C into the first equation:

$T_F = \frac{9}{5}(T_K - 273.15) + 32$,

$\boxed{T_F = \frac{9}{5}T_K - 459.67}$.

11.9 (a) Call the new units °N. Then, 10 N° = 28 F°. The zero of the new scale is at 70°F, so

$$\boxed{T(°N) = \frac{5}{14}\left(T(°F) - 70\right)}.$$

(b) Insert $T(°F) = \frac{9}{5}T(°C) + 32$ to get

$$T(°N) = \frac{5}{14}\left(\frac{9}{5}T(°C) + 32 - 70\right) = \frac{9}{14}T(°C) - \frac{95}{7},$$

$$\boxed{T(°N) = \frac{9}{14}T(°C) - 13.6}.$$

11.11 Using Eq.(11.4) we find that the diameter of the iron tire at temperature T is $L = L_o[1 + \alpha(T - T_o)]$. Inserting the numerical values gives
$L = 1.50\text{ m}[1 + (12 \times 10^{-6}\,°C^{-1})(800°C - 15°C)] = \boxed{1.51\text{ m}}$.

11.13 The amount lost to overflow is the amount of the expansion, assuming that the volume of the tank is unchanged. The additional volume of gasoline is given by $\Delta V = \beta V_o\,\Delta T$, where β is the volume thermal expansion coefficient. Since β is given in °C, we need to find ΔT in °C: $\Delta T_C = \frac{5}{9}\Delta T_F$. Thus,
$\Delta V = (950 \times 10^{-6}\,°C^{-1})(60\text{ L})\left(\frac{5°C}{9°F}\,50°F\right) = \boxed{1.6\text{ L}}$.

11.15 The increase in height is ΔL. From Eq.(11.3) we find
$\Delta L = \alpha L_o\,\Delta T = (12 \times 10^{-6}\,°C^{-1})(50\text{ m})(35\text{ C°}) = 0.021\text{ m} = \boxed{2.1\text{ cm}}$.

11.17 Length of steel bar changes by
$\Delta L_s = L_o\alpha\Delta T = 1.0\text{ m}(12 \times 10^{-6}\,°C^{-1})(100\text{ °C}) = 1.2\text{ mm}$.
The copper bar changes by
$\Delta L_c = L_o\alpha\Delta T = 0.50\text{ m}(17 \times 10^{-6}\,°C^{-1})(100\text{ °C}) = 0.85\text{ mm}$.
So the top of the copper rod moves up by 1.2 mm while its length increases by 0.85 mm. Thus the bottom of the copper rod moves by the difference,
$1.2\text{ mm} - 0.85\text{ mm} \approx \boxed{0.4\text{ mm up}}$.

11.19 (a) The diameter of the ball at 275°C can be computed from the linear expansion equation (Eq.11.4). Let L_b and L_{bo} be the diameter of the ball at 275°C and 15°C, respectively. Then the diameter of the ball at 275°C is
$L_b = L_{bo}[1 + \alpha(T - T_o)] = 1.998\text{ cm}\,[1 + (17 \times 10^{-6}\,°C^{-1})(275°C - 15°C)]$,
$L_b = 2.007$ cm. $\boxed{\text{No}}$, the ball will no longer pass through the hole in the plate.
(b) The diameter of the hole increases with temperature. It is given by
$L_p = L_o[1 + \alpha(T - T_o)]$,
$L_p = 2.005\text{ cm}[1 + (12 \times 10^{-6}\,°C^{-1})(275°C - 15°C)] = 2.011\text{ cm}$.
$\boxed{\text{Yes}}$, the ball passes through the hole when both ball and plate are at 275°C.

11.21 For simplicity assume a rectangular plate with dimensions L_x and L_y. The area of the plate is given by
$A = L_xL_y = L_{xo}\big(1 + \alpha(T - T_o)\big)L_{yo}\big(1 + \alpha(T - T_o)\big)$,
$A = L_{xo}L_{yo}\big(1 + 2\alpha(T - T_o) + \alpha^2(T - T_o)^2\big)$.
To first order in small quantities we can ignore the term in α^2. Let $A_o = L_{xo}L_{yo}$.
Then the area becomes $\boxed{A = A_o\big(1 + 2\alpha(T - T_o)\big)}$.

11.23 (a) We need to know the energy required to raise 1.0 kg of water by 100°C - 22°C = 78°C. Since 1 kcal will raise 1 kg by 1°C, it will require 78 kcal to raise the 1.0 kg of water to the boiling point of 100°C.Thus

$Q = 78$ kcal = (78 kcal)(1000 cal/kcal)(4.187 J/cal) = $\boxed{3.3 \times 10^5 \text{ J}}$.

(b) Set the work done (mgh) equal to the energy Q.

$mgh = Q$.

$h = Q/mg = 3.3 \times 10^5 \text{ J}/(50 \text{ kg} \times 9.8 \text{ m/s}^2) = \boxed{670 \text{ m}}$.

11.25 Assume that all of the potential energy goes into changing the temperature of the water. For one gram of water to fall from top to bottom the energy available is

$Q = mgh = (1.0 \times 10^{-3} \text{ kg})(9.81 \text{ m/s}^2)(23 \text{ m}) = 0.226$ J.

Converting to calories we get

$0.226 \text{ J} = 0.226 \text{ J} \times \frac{1 \text{ cal}}{4.187 \text{ J}} = 0.054$ cal. This amount of energy would raise the temperature of 1 gram of water by $\boxed{0.054°\text{C}}$.

11.27 To raise 1 kg of water 1°C requires 1000 cal = 4187 J.

The 1500-W heater is releasing 1500 J/s to the water. In one second there is enough energy to raise 1 kg of water by $\frac{1500 \text{ J}}{4187 \text{ J/°C}} = 0.358$ °C.

Thus the temperature of the water rises at approximately $\boxed{0.36°\text{C/s}}$.

11.29 The initial potential energy is converted to thermal energy:

$mgh = Q = m_w c\Delta T$, where m_w is the mass of the water and c is its specific heat.

$$\Delta T = \frac{mgh}{m_w c} = \frac{(10 \text{ kg})(9.8 \text{ m/s}^2)(1.6 \text{ m})}{(2.0 \text{ kg})(4187 \text{ J/kg·°C})} = \boxed{0.019°\text{C}}.$$

11.31 Heat lost plus heat gained = 0. $Q = mc\Delta T$.

$m_1c(T - T_{1o}) + m_2c(T - T_{2o}) = 0$.

(250 g)(c)(T – 85°C) + (80g)(c)(T – 15°C) = 0,

T(250 g + 80 g) = (250 g)(85°C) + (80 g)(15°C),

$$T = \frac{21250 + 1200}{250 + 80} °\text{C} = \boxed{68°\text{C}}.$$

11.33 Heat lost plus heat gained = 0. $Q = mc\Delta T$.

$m_1c_1(T - T_{1o}) + m_2c_2(T - T_{2o}) = 0$.

The specific heats are found in Table 11.3.

(150 g)(1.00 cal/g·°C)(T – 86°C) + (200 g)(0.200 cal/g·°C)(T – 22°C) = 0,

150 T – 12900°C + 40 T – 880°C = 0,

190 T – 13780°C = 0,

$$T = \frac{13780°\text{C}}{190} = \boxed{73°\text{C}}.$$

11.35 Heat lost plus heat gained = 0. $Q = mc\Delta T$.

$m_1c_1(T - T_{1o}) + m_2c_2(T - T_{2o}) = 0$.

(1.0 kg)(30.5 cal/kg·°C)(T – 150°C) + (1.0 kg)(1000 cal/g·°C)(T – 23°C) = 0,

30.5 T – 4575°C + 1000 T – 23000°C = 0,

$1030.5\ T - 27575°C = 0,$

$$T = \frac{27575°C}{1030.5} = 26.76°C \approx \boxed{27°C}.$$

11.37 This problem may be solved with one energy conservation equation by considering all three parts of water. We simply add the heat change ΔQ for each of the three masses of water and set the total equal to zero.

$\Delta Q_1 + \Delta Q_2 + \Delta Q_3 = 0.$

$m_1 c \Delta T_1 + m_2 c \Delta T_2 + m_3 c \Delta T_3 = 0.$

Since all three constituents are water, they have the same specific heat c, so that factor can be divided out of the equation.

$250\ g\ (30°C - 80°C) + 180\ g\ (30°C - 10°C) + 300\ g\ (30°C - T) = 0,$

$300\ T = -12500°C + 3600°C + 9000°C = 100°C,$

$T = 100°C/300 = \boxed{0.33°C}.$

11.39 Estimate: $m_{water} \approx 200$ g, $m_{ice} \approx 35$ g, ignore effects of the cup.

$(200\ g)(1.00\ cal/g°C)(T - 100°C)+(35\ g)[79.9\ cal/g+(1.00\ cal/g°C)(T - 0°C)] = 0,$

$200\ T - 20000°C + 2796.5°C + 35\ T, \quad T \approx \boxed{73°C}.$

11.41 $m_{ice}L_f = 1.00\ g_{steam}L_v,$

$m_{ice} = 1.00\ g\ (539\ cal/g)/(79.9\ cal/g) = \boxed{6.75\ g}.$

11.43 Heat released in condensing 20 g of steam is

$Q = mL_v = (20\ g)(539\ cal/g) = 10780\ cal.$

Energy needed to raise 80 g of water from 20°C to 100°C is

$Q = mc\Delta T = (80\ g)(1.00\ cal/g°C)(80°C) = 6400\ cal.$

Since the energy available from the steam exceeds the energy needed to heat the water, the final mixture will come to a final temperature of $\boxed{100°C}$.

11.45 Energy is supplied at the rate of 30 W = 30 J/s. Heat of fusion of ice is $L_f = 3.34 \times 10^5$ J/kg.

$$\text{Rate of ice melting} = \frac{\text{power}}{L_f} = \frac{30\ J/s}{3.34 \times 10^5\ J/kg} = 9.0 \times 10^{-5}\ kg/s,$$

Rate = 0.090 g/s. Since one g of water occupies 1 cm^3, the volume of water produced per second is 0.090 cm^3/s. The volume of water produced per minute is $(0.090\ cm^3/s)(60\ s/min) = \boxed{5.4\ cm^3/min}$.

11.47 Total energy needed for:

melting ice = $300\ g \times 79.7\ cal/g = 23910$ cal,

heat melted ice to 100° = $300\ g \times 1\ cal/g°C \times 100°C = 30000$ cal,

heat water = $500\ g \times 1\ cal/g°C \times 80°C = 40000$ cal,

boil all water = $800\ g \times 539\ cal/g = 431200$ cal,

total energy needed is 5.25×10^5 cal $\times$ 4.187 J/cal = 2.20×10^6 J.
Power is energy/time, so the time required is the energy divided by the power.
$t = Q/P = 2.20 \times 10^6 \text{ J}/50 \text{ W} = \boxed{4.40 \times 10^4 \text{ s}} = 12$ h 13 min.

11.49 We can calculate the rate of heat loss from Eq.(11.8)

$$\frac{\Delta Q}{\Delta t} = KA\frac{T_2 - T_1}{L},$$

where K = 0.046 W/m·°C, A = 1 m × 1.5 m, and L = 0.15 m.

$$\frac{\Delta Q}{\Delta t} = (0.046 \text{ W/m}\cdot°\text{C})(1.5 \text{ m}^2)\frac{20°\text{C}}{0.15 \text{ m}} = \boxed{9.2 \text{ W}}.$$

11.51 The rate of energy transfer is given by Eq.(11.8).

$$\frac{\Delta Q}{\Delta t} = KA\frac{T_2 - T_1}{L} = (398 \text{ W/m}\cdot°\text{C})(1.0 \times 10^{-4} \text{ m}^2)\frac{100°\text{C}}{1.0 \text{ m}},$$

$$\frac{\Delta Q}{\Delta t} = \boxed{4.0 \text{ W}}.$$

11.53 From Eq.(11.10), for T = 300 K we get

$P/A = \sigma e T^4 = (5.67 \times 10^{-8} \text{ W}\cdot\text{m}^{-2}\cdot\text{K}^{-4})(1.00)(300 \text{ K})^4 = \boxed{459 \text{ W/m}^2}$.

For T = 1000 K we get

$P/A = (5.67 \times 10^{-8} \text{ W}\cdot\text{m}^{-2}\cdot\text{K}^{-4})(1.00)(1000 \text{ K})^4 = \boxed{5.67 \times 10^4 \text{ W/m}^2}$.

For T = 3000 K we get

$P/A = (5.67 \times 10^{-8} \text{ W}\cdot\text{m}^{-2}\cdot\text{K}^{-4})(1.00)(3000 \text{ K})^4 = \boxed{4.59 \times 10^6 \text{ W/m}^2}$.

11.55 Use the Stefan-Boltzmann law: $P = \sigma e A T^4$. Upon rearranging we find

$$T = \left(\frac{P/A}{\sigma e}\right)^{1/4}. \qquad \text{For } e = 1.00,\ T = \left(\frac{P/A}{\sigma}\right)^{1/4}.$$

$$T = \left(\frac{6.25 \times 10^7 \text{ W/m}^2}{5.67 \times 10^{-8} \text{ W/(m}^2\cdot\text{K}^4)}\right)^{1/4} = \boxed{5760 \text{ K}}.$$

11.57 $R = L/K$, so $K/L = 1/R$. Note that R values are not given in SI units.

(a) $$\frac{\Delta Q}{\Delta t} = \frac{A}{R}(T_2 - T_1).$$

For this problem it is easiest to use the British units to find

$$\frac{\Delta Q}{\Delta t} = \frac{8.00 \text{ ft} \times 16.4 \text{ ft}}{15 \text{ ft}^2\text{h}\cdot°\text{F/Btu}}\ 25°\text{C} \times \frac{9 \text{ F}°}{5 \text{ C}°} = 393.6 \text{ Btu/h}.$$

Now convert to SI units.

$$\frac{\Delta Q}{\Delta t} = 393.6 \text{ Btu/h} \times \frac{1\text{h}}{3600 \text{ s}} \times 1060 \text{ J/Btu} = \boxed{116 \text{ W}}.$$

(b) Reduce area with R of 15 and add a contribution from the area of window with R = 1.

$$\frac{\Delta Q_{wall}}{\Delta t} = \frac{A}{R}(T_2 - T_1) = \frac{120.4\ \text{ft}^2}{15\ \text{ft}^2\text{h}\cdot{}^\circ\text{F/Btu}}\left(25^\circ\text{C}\ \frac{9\ {}^\circ\text{C}}{5^\circ\text{F}}\right) = 361.3\ \text{Btu/h}.$$

$$\frac{\Delta Q_{window}}{\Delta t} = \frac{A}{R}(T_2 - T_1) = \frac{10.76\ \text{ft}^2}{1\ \text{ft}^2\text{h}\cdot{}^\circ\text{F/Btu}}\left(25^\circ\text{C}\ \frac{9\ {}^\circ\text{C}}{5^\circ\text{F}}\right) = 484.2\ \text{Btu/h}.$$

$$\frac{\Delta Q_{total}}{\Delta t} = 845.5\ \text{Btu/h} \times \frac{1\text{h}}{3600\ \text{s}} \times 1060\ \text{J/Btu} = \boxed{249\ \text{W}}.$$

11.59 (a) When the temperature reaches a steady state, the heat flow through the rod is uniform.

$$\frac{\Delta Q}{\Delta t} = K_{iron}A\frac{T - 0^\circ\text{C}}{0.50\ \text{m}} = K_{copper}A\frac{100^\circ\text{C} - T}{0.50\ \text{m}}.$$

Upon factoring out the lengths (0.50 m) and the area A we find that

$K_{iron}T = K_{copper}100^\circ\text{C} - K_{copper}T$.

$$T = \frac{K_{copper}}{K_{iron} + K_{copper}}100\ {}^\circ\text{C} = \frac{398}{80.3 + 398}100^\circ\text{C} = \boxed{83.3^\circ\text{C}}.$$

(b) $$\frac{\Delta Q}{\Delta t} = K_{iron}A\frac{\Delta T}{L} = (80.3\ \text{W/m}^\circ\text{C})(1.5 \times 10^{-4}\ \text{m}^2)\frac{83.3^\circ\text{C}}{0.50\ \text{m}},$$

$$\frac{\Delta Q}{\Delta t} = \boxed{2.0\ \text{W}}.$$

11.61 The increase in volume is given by

$\Delta V = \beta V_o \Delta T$, where $\beta = 3\alpha$ and $V = \frac{4}{3}\pi r_o^3$.

$$\Delta V = (3 \times 24 \times 10^{-6}{}^\circ\text{C}^{-1})\left(\frac{4}{3}\right)\pi(10.0\ \text{cm})^3(100^\circ\text{C} - 0^\circ\text{C}) = \boxed{30\ \text{cm}^3}.$$

11.63 (a) The brakes must dissipate an energy equal to the original kinetic energy of the car. $W = \frac{1}{2}mv^2 = \frac{1}{2}(1900\ \text{kg})(30\ \text{m/s})^2 = \boxed{8.6 \times 10^5\ \text{J}}$.

(b) $P = W/t = 8.55 \times 10^{-5}\ \text{J}/15\ \text{s} = \boxed{5.7 \times 10^4\ \text{W}}$.

(c) $W = (mc)\Delta T$.

$$\Delta T = \frac{W}{mc} = \frac{8.6 \times 10^5\ \text{J/s}}{(0.75\ \text{kcal/}^\circ\text{C})(4190\ \text{J/kcal})} = \boxed{270^\circ\text{C}}.$$

11.65 $$\frac{\Delta Q}{\Delta t} = KA\frac{\Delta T}{L} = \frac{A\,\Delta T}{R}.$$

$$R = \frac{A\,\Delta T}{\Delta Q/\Delta t} = \frac{\Delta T}{\Delta Q/A\Delta t} = \frac{20^\circ\text{C}}{10\ \text{W/m}^2} = 2\ \text{m}^2\cdot\text{s}\cdot{}^\circ\text{C/J}.$$

$R = (2\ m^2 \cdot s \cdot °C/J)(3.28\ ft/m)^2(1\ h/3600\ s)(9\ F°/5\ C°)(1060\ J/Btu),$

$R = \boxed{11.4\ ft^2 \cdot h \cdot °F/Btu}.$

11.67 Note, 1 L of water = 1000 g = 1.000 kg.

$$\frac{\Delta Q}{\Delta t} = 85\ W = \frac{\Delta m}{\Delta t} c\Delta T.$$

Here $\Delta m = 1.00$ kg and $\Delta t = 5$ min = 300 s.

$$\frac{\Delta Q}{\Delta t} = \left(\frac{1.00\ kg}{300\ s}\right) 4190\ J/kg°C\ \Delta T.$$

$$\Delta T = \frac{85 \times 300}{1.00 \times 4190} °C = \boxed{6.1°C}.$$

11.69 Calculate the energy to raise each gram of ice to 20°C and compare it with the energy released in cooling a gram of steam to 20°C.
For the ice: Q/g = 79.7 cal + 20 cal = 99.7 cal.
For the steam: Q/g = 539 cal + 80 cal = 619 cal.

$m_{steam}(Q/g)_{steam} = 300\ g_{ice}\ (Q/g)_{ice},$

$$m_{steam} = \frac{(300\ g)(99.7\ cal/g)}{619\ cal/g} = \boxed{48.3\ g}.$$

11.71 (a) Fractional change $= \dfrac{L - L_o}{L_o} = \alpha(T - T_o).$

Fractional change $= 1.85 \times 10^{-6}/°C \times (38°C - 20°C) = \boxed{3.33 \times 10^{-4}}.$

(b) $T_o = 2\pi\sqrt{L_o/g}, \qquad T = 2\pi\sqrt{L/g}.$

$$\frac{T}{T_o} = \sqrt{\frac{L}{L_o}} = \sqrt{1.0000 + (3.33 \times 10^{-4})} = 1.0000 + (1.67 \times 10^{-4}).$$

We have used the binomial theorem to get this result. That is for small x,

$(1 + x)^{1/2} = 1 + \frac{1}{2}x + \ldots \approx 1 + \frac{1}{2}x.$

$T = T_o\left[1.0000 + (1.67 \times 10^{-4})\right],$

$T - T_o = T_o(1.67 \times 10^{-4}),$

$T = (24\ h)(3600\ s/h)(1.67 \times 10^{-4}) = \boxed{14.4\ s}.$

11.73 $\dfrac{\Delta Q}{\Delta t} = KA\dfrac{\Delta T}{L} = A\dfrac{\Delta T}{R}.$ For a piece of insulation of R_1 followed by insulation of R_2, the heat flow through each is the same. Let the temperature of the common side be T, the others temperatures T_H and T_C. Then

$\dfrac{T_H - T}{R_1} = \dfrac{T - T_C}{R_2}.$ From this we find $T = \dfrac{R_1 T_C + R_2 T_H}{R_1 + R_2}.$

$$\frac{\Delta Q}{\Delta t} = A\frac{T_H - T}{R_1} = \frac{A}{R_1}\left(T_H - \frac{R_1 T_C + R_2 T_H}{R_1 + R_2}\right)$$

$$\frac{\Delta Q}{\Delta t} = \frac{A}{R_1}\left(\frac{T_H R_1 + T_H R_2 - R_1 T_C - R_2 T_H}{R_1 + R_2}\right) = A\frac{T_H - T_C}{R_1 + R_2}.$$

So the effective value of R is $\boxed{R = R_1 + R_2}$.

11.75 The aluminum block has a mass $m = \rho V$ where $\rho = 2.7\ \text{g/cm}^3$ and $V = (4.0\ \text{cm})^3$. So, $m = 2.7\ \text{g/cm}^3 \times 64\ \text{cm}^3 = 172.8\ \text{g}$.

The interior volume of the insulating container is $V_o = (10\ \text{cm})^3 = 1000\ \text{cm}^3$.

The volume of the 300 g of water is $300\ \text{cm}^3$.

The available volume for additional water is the container volume less the volume of the aluminum block and the volume of water already present:

$V_{add} = 1000\ \text{cm}^3 - 64\ \text{cm}^3 - 300\ \text{cm}^3 = 636\ \text{cm}^3$. Consequently, the most water that can be added without overflow is 636 g.

Now consider the equation for conservation of heat energy.

$\Delta Q_{Al} + \Delta Q_{water\ initial} + \Delta Q_{water\ added} = 0.$

$m_{Al}c_{Al}\Delta T_{Al} + m_{water\ init}c_{water}\Delta T_{water\ init} + m_{water\ add}c_{water}\Delta T_{water\ add} = 0.$

Assume that you can cool to 12°C and find out how much water is needed if the water is at the coldest temperature allowed of 0°C.

$(172.8\ \text{g})(0.215\ \text{cal/g}\cdot°\text{C})(12°\text{C} - 95°\text{C}) + (300\ \text{g})(1.0\ \text{cal/g}\cdot°\text{C})(12°\text{C} - 30°\text{C})$

$\quad + m(1.0\ \text{cal/g}\cdot°\text{C})(12°\text{C} - 0°\text{C}) = 0,$

$-3084\ \text{cal} - 5400\ \text{cal} + m\ (12\ \text{cal/g}) = 0,$

$$m = \frac{8484\ \text{cal}}{12\ \text{cal/g}} = 707\ \text{g}.$$

However, we found earlier that because of the limited available space for added water we could only add 636 g of water. So we cannot cool the system to 12°C within the constraints of the problem. The answer is $\boxed{\text{no.}}$

Chapter 12

GAS LAWS AND KINETIC THEORY

12.1 The liquid will reach a height for which the pressure at the bottom will equal the pressure of the atmosphere: $P = \rho gh$. Thus,

$$h = \frac{P}{\rho g} = \frac{101 \text{ kPa}}{(1.26 \times 10^3 \text{ kg/m}^3)(9.81 \text{ m/s}^2)} = \boxed{8.17 \text{ m}}.$$

12.3 The force is $F = PA = (1.01 \times 10^5 \text{ N})(0.21 \text{ m} \times 0.28 \text{ m}) = \boxed{5.9 \times 10^3 \text{ N}}$. There is an equal pressure from below.

12.5 1 atm = 101 kPa will support 10.4 m of water.
0.20 in of water = 0.20 in × 2.54 cm/in = 0.508 cm.
The pressure due to 0.508 cm of water is

$$0.508 \times 10^{-2} \text{ m} \frac{1.01 \times 10^5 \text{ Pa}}{10.4 \text{ m}} = 49.3 \text{ N/m}^2.$$

The force on the window is the product of the pressure with the area.
$F = PA = 49.3 \text{ N/m}^2 (3.0 \text{ m} \times 4.0 \text{ m}) = \boxed{590 \text{ N}}$.

12.7 Use Boyle's law: $PV = P_oV_o$, so $P = P_o \frac{V_o}{V_o/3} = 3 P_o = \boxed{3 \text{ atm}}$.

12.9 Apply Boyle's law. $P_1V_1 = P_2V_2$, where 1 refers to conditions at the bottom and 2 conditions at the surface.
Since $V_2 = 2 V_1$, $P_1 = 2 P_2 = 2(1 \text{ atm}) = 2$ atm.
The pressure at the bottom is given by
$P_1 = P_{atm} + \rho gh = 1 \text{ atm} + \rho gh$.
So $\rho gh = 1 \text{ atm} = 1.01 \times 10^5 \text{ N/m}^2$.

$$h = \frac{1.01 \times 10^5 \text{ N/m}^2}{(1000 \text{ kg/m}^3)(9.81 \text{ m/s}^2)} = \boxed{10.3 \text{ m}}.$$

12.11 For constant pressure V/T is constant. Thus

$$\frac{V_1}{T_1} = \frac{V_2}{T_2}. \quad \text{So,} \quad V_2 = V_1 \frac{T_2}{T_1} = 1.00 \text{ L} \frac{(273 + 4) \text{ K}}{(273 + 23) \text{ K}} = \boxed{0.94 \text{ L}}.$$

12.13 Length is proportional to the volume since the cross-section is constant.
Law of Charles becomes L/T = constant. So, $L = L_o \frac{T}{T_o}$.

$$L = 50 \text{ cm} \frac{373}{273} = \boxed{68 \text{ cm from the closed end}}.$$

12.15 Length is proportional to the volume since the cross-section is constant.
Law of Charles becomes L/T = constant. $T = -10°C = 263$ K.

$$\frac{T_o}{L_o} = \frac{T}{L},$$

$$T_o = T\frac{L_o}{L} = 263\text{ K}\frac{50\text{ cm}}{42\text{ cm}} = 313\text{ K} = \boxed{40\text{ °C}}.$$

12.17 (a) PV/T is the same for both so the number of molecules is the same. The ratio of hydrogen to helium molecules is $\boxed{\text{one}}$.

(b) The piston will not move due to change in T. It will be in the $\boxed{\text{same place}}$.

12.19 $PV = nRT$. $V = \frac{nRT}{P}$. From Table 12.1 we find that CO_2 has a molecular mass of 44 g/mol. Thus the number of moles in 20 g is $n = \frac{20\text{ g}}{44\text{g/mol}}$. The temperature in kelvins is $T = T_C + 273 = 298$ K.

$$V = \frac{(20\text{ g})(8.314\text{ J/mol K})(298\text{ K})}{(44\text{ g/mol})(1.01 \times 10^5\text{ N/m}^2)} = 1.1 \times 10^{-2}\text{ m}^3 = \boxed{11\text{ L}}.$$

12.21 (a) At STP one cubic meter contains n moles. From the ideal gas law we get

$$n = \frac{PV}{RT} = \frac{1.01 \times 10^5\text{ Pa} \times 1.00\text{ m}^3}{(8.314\text{ J/mol})(273\text{ K})} = 44.5\text{ mol}.$$

The mass per mol is $\frac{1784\text{ g}}{44.5\text{ mol}} = \boxed{40.1\text{ g/mol}}$.

(b) The gas is $\boxed{\text{argon}}$.

12.23 Average value is $\frac{1+3+7+8}{4} = \boxed{4.75}$.

Rms value is $\sqrt{\frac{1^2+3^2+7^2+8^2}{4}} = \boxed{5.55}$.

12.25 (a) The rms speed is $v_{rms} = \sqrt{\frac{3kT}{m}}$.

Square both sides and rearrange the equation to get

$$T = \frac{m}{3k}v_{rms}^2 = \frac{3.34 \times 10^{-27}\text{ kg}}{3 \times 1.38 \times 10^{-23}\text{ J/K}}(2200\text{ m/s})^2 = \boxed{390\text{ K}}.$$

(b) Since the temperature is related to the square of the speed, reducing the speed by one half reduces the temperature by one fourth, $T = \boxed{98\text{ K}}$.

12.27 $KE = \frac{3}{2}kT = \frac{3}{2}(1.38 \times 10^{-23}\text{ J/K})(300\text{ K}) = \boxed{6.21 \times 10^{-21}\text{ J}}$.

12.29 The internal energy is defined as $U = \frac{3}{2}nRT$,

$$U = \frac{3}{2}(1\text{ mol})(8.314\text{ J/mol·K})(300\text{ K}) = \boxed{3740\text{ J}}.$$

12.31 (a) Since the speed of sound is proportional to the rms speed, consider the form of $v_{rms} = \sqrt{\frac{3kT}{m}}$.

Thus the speed of sound is proportional to $\boxed{\sqrt{\frac{3kT}{m}}}$.

12.33 The rms speed is $v_{rms} = \sqrt{\frac{3kT}{m}}$.

$$\frac{v_{rms}(He)}{v_{rms}(H_2)} = \frac{\sqrt{\frac{3kT}{m(He)}}}{\sqrt{\frac{3kT}{m(H_2)}}} = \sqrt{\frac{m(H_2)}{m(He)}} = \sqrt{\frac{1}{2}} = \boxed{0.71}.$$

12.35 Start with $v_{rms} = \sqrt{\frac{3P}{\rho}}$. The density $\rho = \frac{M}{V}$, where the total mass $M = Nm$.

Making the substitution gives

$v_{rms} = \sqrt{\frac{3PV}{Nm}}$. From the ideal gas law $PV = nRT = \frac{NRT}{N_A} = NkT$.

Thus $v_{rms} = \sqrt{\frac{3NkT}{Nm}} = \boxed{\sqrt{\frac{3kT}{m}}}$.

12.37 Graph the quantity $n = n_o e^{-mgz/kT}$ for $n_o = 1$, $T = 23°C = 300$ K, and $m = (32 \text{ g/mol})/(6.02 \times 10^{23}/\text{mol}) = 5.32 \times 10^{-23}$ g $= 5.32 \times 10^{-26}$ kg.

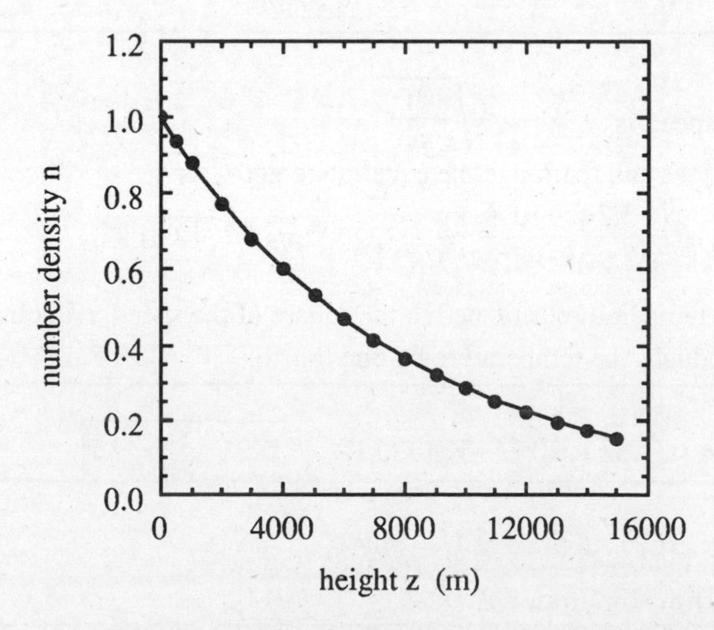

The graph can also be made using logarithmic coordinates.

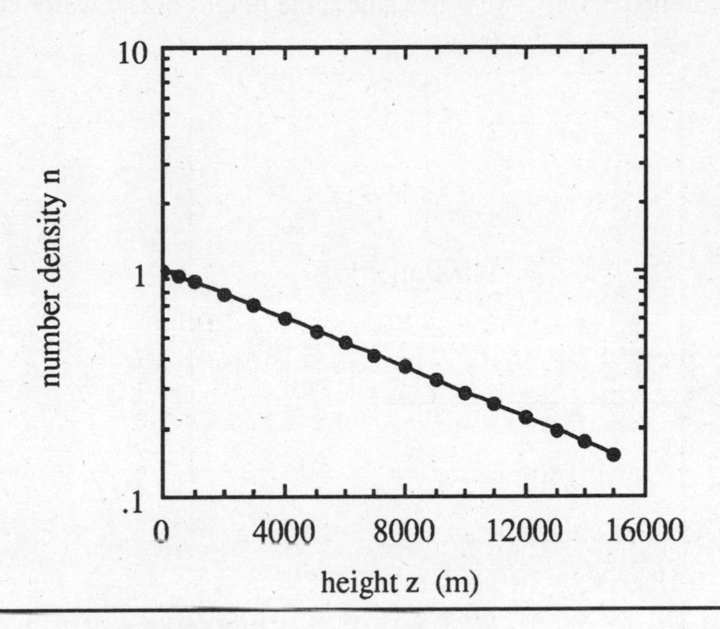

12.39 $v_{sound} = \sqrt{\dfrac{1.4\,P}{\rho}}$ and $v_{rms} = \sqrt{\dfrac{3\,P}{\rho}}$.

$$\frac{v_{rms}}{v_{sound}} = \sqrt{\frac{3}{1.4}} = \boxed{1.46}.$$

12.41 Mass of $N_2 = 4.67 \times 10^{-26}$ kg, $P = 1.01 \times 10^5$ Pa, $T = 300$ K.

$P = \frac{1}{2} P_o = P_o\, e^{-mgz/kT}$.

Take the logarithm of both sides to get

$-mgz/kT = \ln\left(\frac{1}{2}\right) = -0.693$,

$$z = \frac{0.693\,kT}{mg} = \frac{0.693(1.38 \times 10^{-23}\ \text{J/K})(300\ \text{K})}{(4.67 \times 10^{-26}\ \text{kg})(9.81\ \text{m/s}^2)} = \boxed{6270\ \text{m}}.$$

12.43 Decreasing volume by 20% means that the new volume becomes 0.80 of the initial volume. Then applying Boyle's law:

$$P = P_o \frac{V_o}{V} = 1.01 \times 10^5\ \text{Pa}\ \frac{V_o}{0.80\ V_o} = \boxed{1.26 \times 10^5\ \text{Pa}}.$$

12.45 (a) From the ideal gas law we get $n = \dfrac{PV}{RT}$.

$$n = \frac{15(1.01 \times 10^5\ \text{Pa})(1.0 \times 10^{-3}\ \text{m}^3)}{8.314\ \text{J/mol}\cdot\text{K}(291\ \text{K})} = \boxed{0.63\ \text{mol}}.$$

(b) Molecular mass is 28 g/mol, so the mass within the container is

$m = (28\ \text{g/mol})(0.63\ \text{mol}) = \boxed{18\ \text{g}}$.

12.47 We will ignore the thickness of the piston. When the water is added, the gas is compressed so that the piston moves down a distance y. The height of the enclosed region is 0.500 m – y. When filled, the height of the water column is 0.500 m + y, as shown in the figure.

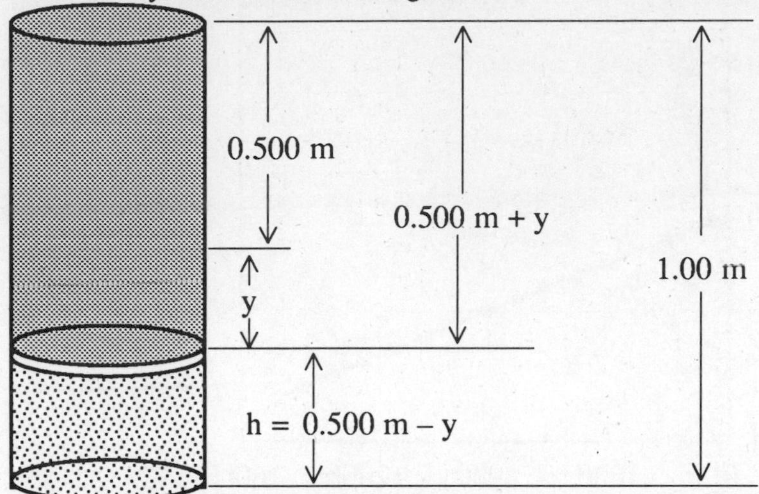

$P_1 = 1.01 \times 10^5$ Pa, $V_1 = A(0.500 \text{ m})$,
$V_2 = A(0.500 \text{ m} - y)$ and $P_2 = 1.01 \times 10^5 \text{ Pa} + \rho g(0.500 \text{ m} + y)$,
$P_1V_1 = P_2V_2$, Put all quantities in SI units to get,
$1.01 \times 10^5 (0.500)A = [1.01 \times 10^5 + 1000(9.8)(0.500 + y)](0.500 - y)A$,
$5.05 \times 10^4 = (1.059 \times 10^5 + 9800y)(0.500 - y)$,
$-9800\, y^2 - 1.010 \times 10^5\, y + 5.295 \times 10^4 = 5.05 \times 10^4$,
$y^2 + 10.30\, y - 0.250 = 0$.
Solve this quadratic equation to get
$y = 0.024$. $h = 0.500 - 0.024 = 0.476$ m.
Ratio of h to total length = 0.476 m/1.00 m = $\boxed{0.476}$.

12.49 (a) $PV = nRT$, so $N = nN_A = \dfrac{PVN_A}{RT}$

$$N = \frac{(1.01 \times 10^5 \text{ Pa})(1.0 \text{ m}^3)(6.02 \times 10^{23}/\text{mol})}{(8.314 \text{ J/K})(300 \text{ K})} = \boxed{2.44 \times 10^{25}}.$$

(b) $PV = \frac{3}{2} N\, \overline{KE}$.

$KE_{total} = N\,\overline{KE} = \frac{2}{3} PV = \frac{3}{2}\, 1.01 \times 10^5 \text{ N/m}^2 \times 1 \text{ m}^3 = \boxed{1.5 \times 10^5 \text{ J}}$.

(c) $\frac{1}{2} mv^2 = \frac{1}{2} (0.17 \text{ kg})\, v^2 = 1.5 \times 10^5$ J.

$$v = \sqrt{\frac{2 \times 1.5 \times 10^5}{0.17}} \text{ m/s} = \boxed{1.34 \times 10^3 \text{ m/s}}.$$

12.51 (a) $v_{rms} = \sqrt{\dfrac{3kT}{m}}$, $v_{escape} = \sqrt{\dfrac{2GM_E}{R_E}} = \sqrt{2gR_E}$. Set $v_{rms} = v_{escape}$,

then $\dfrac{3kT}{m} = 2gR_E$. For nitrogen, $m = 28 \times N_A = 4.67 \times 10^{-26}$ kg.

$$T = 2gR_Em/3k = \frac{2(9.8\text{ m/s}^2)(6.38 \times 10^6\text{ m})(4.67 \times 10^{-26}\text{ kg})}{3(1.38 \times 10^{-23}\text{ J/K})}.$$

For nitrogen, T = $\boxed{1.41 \times 10^5\text{ K}}$.

(b) For oxygen, $m = 32 \times N_A$, and T = $\boxed{1.60 \times 10^5\text{ K}}$.

(c) $\boxed{\text{Ratio of nitrogen to oxygen is slightly smaller.}}$

A12.1 From the graph and by calculation we get a slope of $\boxed{-0.061}$.

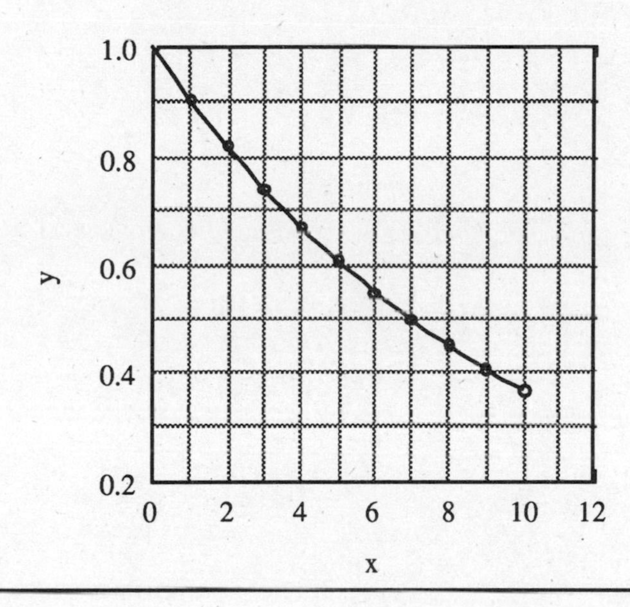

A12.3 A graph of $y = e^{-x}$.

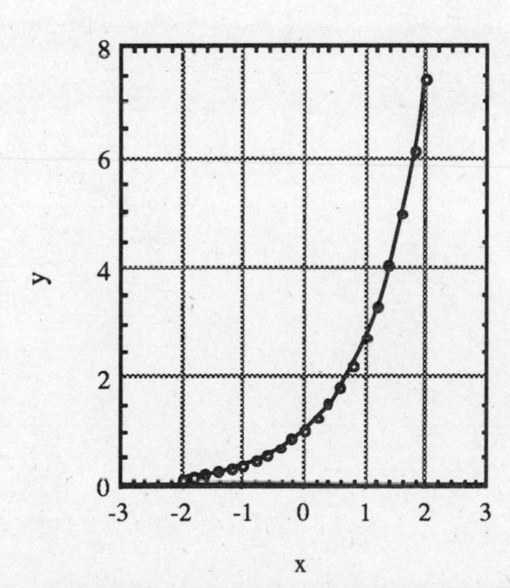

A12.5 $R = R_o e^{-0.693t/T} = (1200\text{ particles/min})\ e^{-0.693(120\text{ min}/26\text{ min})}$,
$R = (1200\text{ particles/min})(0.0408) = \boxed{49\text{ particles/min}}$.

A12.7 $e^x = 250 \times 10^6$, $x = \ln(250 \times 10^6) = \boxed{19.34}$.

Chapter 13

THERMODYNAMICS

13.1 Use the first law of thermodynamics.

$\Delta U = Q - W = 5.00 \text{ kcal} \times 4187 \text{ J/kcal} - (-2000 \text{ J}) = \boxed{22.9 \text{ kJ}}$.

13.3 Use the first law of thermodynamics.

$\Delta U = Q - W = 50 \text{ cal} \times 4.187 \text{ J/cal} - 50 \text{ J} = \boxed{160 \text{ J}}$.

13.5 Input energy is $75 \text{ Btu/min} \times 1060 \text{ J/Btu} \times 1 \text{ min}/60 \text{ s} = 1.3 \text{ kW}$.
If the generator produces 1000 W of electrical power, there is still some energy left to the environment as heat. The device is $\boxed{\text{impossible}}$.

13.7 Use the first law of thermodynamics.

$\Delta U = Q - P\Delta V = (10.0 \text{ kcal})(4187 \text{ J/kcal}) - 3.0 \times 10^5 \text{ N/m}^2(0.10 \text{ m}^3)$,

$\Delta U = 41870 \text{ J} - 30000 \text{ J} = \boxed{1.2 \times 10^4 \text{ J}}$.

13.9 Use the first law of thermodynamics.

$\Delta U = Q - W$, but $W = 0$.

$\Delta U = mc\Delta T = (0.010 \text{ kg})(638 \text{ J/kg·K})(60 \text{ K} - 20 \text{ K}) = \boxed{260 \text{ J}}$.

13.11 The work done is $W = \sum \Delta W = \sum P\Delta V$.
For an isobaric process, the pressure is constant so that

$\sum P\Delta V = P\sum \Delta V = P(V_2 - V_1)$. Thus $W = P(V_2 - V_1)$.

13.13 (a) Use the first law of thermodynamics for motion along acb to find ΔU.

$\Delta U = Q - W = 75 \text{ J} - 25 \text{ J} = 50 \text{ J}$.

Along adb, $Q = \Delta U + W = 50 \text{ J} + 10 \text{ J} = \boxed{60 \text{ J}}$.

(b) Along curved path, $Q = \Delta U + W = -50 \text{ J} - 15 \text{ J} = -65 \text{ J}$. The negative sign means that $\boxed{\text{heat is given out}}$.

(c) $\boxed{Q = -65 \text{ J}}$.

13.15 (a) Use the Kelvin relationship. $\dfrac{Q_H}{Q_C} = \dfrac{T_H}{T_C}$.

$\dfrac{Q_H}{17 \text{ J}} = \dfrac{375 + 273}{150 + 273} = \dfrac{648 \text{ K}}{423 \text{ K}}$. $Q_H = 17 \text{ J}(1.53) = \boxed{26 \text{ J}}$.

(b) $W = Q_H - Q_C = 26 \text{ J} - 17 \text{ J} = \boxed{9 \text{ J}}$.

13.17 From Eq.(13.4), the efficiency $= \frac{W}{Q_H} = 1 - T_C/T_H$.

$W = Q_H (1 - T_C/T_H) = 12.0 \text{ J} (1 - 253 \text{ K}/283 \text{ K}) = \boxed{1.27 \text{ J}}$.

13.19 From Eq.(13.4), efficiency $= \frac{T_H - T_C}{T_H} = \frac{293 \text{ K} - 287 \text{ K}}{293 \text{ K}} = \boxed{0.020}$.

13.21 The net work is the area within the PV diagram.

Area = height × width $= (P_o - \frac{1}{2} P_o)(3V_o - V_o) = \boxed{P_o V_o}$.

13.23 From Eq.(13.4), efficiency $= \frac{T_H - T_C}{T_H}$.

$0.30 = \left(1 - \frac{283 \text{ K}}{T_H}\right)$, so $T_H = 283 \text{ K}/0.7 = 404 \text{ K}$.

For 45% efficiency we get $T_H = 283/0.55 = 515$ K, so the temperature of the hot reservoir must be raised by $\boxed{111 \text{ K}}$.

13.25 The work is the area within the PV diagram. The diagram is shown below.

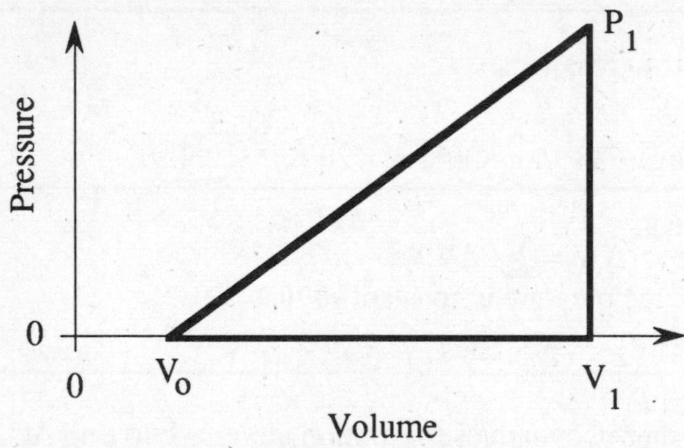

The area enclosed is $\boxed{\frac{1}{2} P_1(V_1 - V_o)}$.

13.27 $\text{efficiency}_{actual} = \frac{W}{Q_H} = \frac{W}{Q_C + W} = \frac{2000 \text{ J}}{2.25(4187 \text{ J}) + 2000 \text{ J}} = \boxed{0.175}$.

$\text{efficiency}_{theoretical} = 1 - \frac{T_C}{T_H} = 1 - \frac{673}{973} = \boxed{0.308}$.

13.29 Theoretical efficiency $= 0.40 = \frac{T_H - T_C}{T_H} = \frac{\Delta T}{T_C + \Delta T}$.

$T_C + \Delta T = \Delta T/0.40$, or $T_C = \Delta T(1/0.40 - 1) = 300 \text{ K}(2.5 - 1)$

$T_C = (300 \text{ K})(1.5) = \boxed{450 \text{ K}}$.

13.31 (a) Theoretical maximum efficiency is

$$\text{efficiency} = 1 - \frac{T_C}{T_H} = 1 - \frac{393}{17273} = \boxed{0.98}.$$

(b) efficiency = W/Q_H where $W/t =$ 22 kW.

Fuel consumption is (1 L/8.5 km)(100 km/3600 s) = 3.27×10^{-3} L/s.

Heat per second is 3.27×10^{-3} l/s $\times 0.35 \times 10^{8}$ J/L = 114 kW.

efficiency = 22 kW/114 kW = $\boxed{0.19}$.

13.33 $$\text{c.p.}_{\text{(refrig.)}} = \frac{T_C}{T_H - T_C} = \frac{273\ \text{K}}{25\ \text{K}} = \boxed{11}.$$

13.35 (a) $$\text{c.p.}_{\text{(heat pump)}} = \frac{T_H}{T_H - T_C} = \frac{294\ \text{K}}{14\ \text{K}} = \boxed{21}.$$

(b) $$\text{c.p.}_{\text{(heat pump)}} = \frac{294\ \text{K}}{33\ \text{K}} = \boxed{8.9}.$$

13.37 Find T_C. From the definition of $\text{c.p.}_{\text{(refrigerator)}}$ we get
$T_C = \text{c.p.}(T_H - T_C)$, so $T_C(\text{c.p.} + 1) = \text{c.p.}(T_H)$.
$T_C = \text{c.p.}\ T_H/(\text{c.p.} + 1) = 5(289\ \text{K})/(5 + 1) = 241\ \text{K} = \boxed{32°\text{C}}$.

13.39 (a) Start with the definition for the efficiency.

$$\text{efficiency} = \frac{T_H - T_C}{T_H} = \frac{420 - 300}{420} = 0.286.$$

But the efficiency is also defined as Q_H/W. So,
$W = (\text{efficiency})(Q_H) = (0.286)(500\ \text{J}) = \boxed{143\ \text{J}}$.

(b) $$\text{c.p.}_{\text{(refrigerator)}} = \frac{T_C}{T_H - T_C} = \frac{300}{420 - 300} = 2.5.$$

c.p. = Q_C/W, so $W = Q_C/\text{c.p.} = 500\ \text{J}/2.5 = \boxed{200\ \text{J}}$.

13.41 $\Delta S = Q/T = mL/T = (100\ \text{g})(334\ \text{J/g})/273\ \text{K} = \boxed{122\ \text{J/K}}$.

13.43 $\Delta S = Q/T = (10.0\ \text{g})(-539\ \text{cal/g})/373\ \text{K} = -14.5\ \text{cal/K} = \boxed{-60.5\ \text{J/K}}$.

13.45 We must compute the entropy change for melting the ice, heating the water, and boiling the water.

$\Delta S = \Delta S_1 + \Delta S_2 + \Delta S_3$

$$\Delta S = \frac{80\ \text{g} \times 79.7\ \text{cal/g}}{273\ \text{K}} + \frac{80\ \text{g} \times 100\ \text{cal/g}}{323\ \text{K}} + \frac{80\ \text{g} \times 539\ \text{cal/g}}{373\ \text{K}}$$

$\Delta S = (23.4 + 24.8 + 115.6)\text{cal/K} = 164\ \text{cal/K} = \boxed{686\ \text{J/K}}$.

13.47 First find the temperature of the mixture.

$m_1c(T - T_1) + m_2c(T - T_2) = 0$. Let (1) be cream and (2) coffee

$(50\text{ g})c(T - 14°\text{C}) + (350\text{ g})c(T - 80°\text{C}) = 0$

Divide by 50 g and by c to get

$T - 14°\text{C} + 7T - 560°\text{C} = 0$. This gives $T = 71.8°\text{C}$.

Average $T_{cream} = (71.8 + 14)/2 = 42.9°\text{C} = 316\text{ K}$ and

average $T_{coffee} = (71.8 + 80)/2 = 75.9°\text{C} = 349\text{ K}$.

$\Delta S = \Delta S_1 + \Delta S_2 = Q_1/T_{1(average)} + Q_2/T_{2(average)}$.

$\Delta S_1 = (50\text{ g})(4.19\text{ J/g})(57.8\text{ K})/316\text{ K} = 38.3\text{ J/K}$,

$\Delta S_2 = (350\text{ g})(4.19\text{ J/g})(-8.2\text{ K})/349\text{ K} = -34.5\text{ J/K}$,

$\Delta S = \boxed{3.8\text{ J/K.}}$

13.49 Heat to warm 1 g ice water = $mc\Delta T = 1\text{ g}(1\text{ cal/g}\cdot°\text{C})\ 37°\text{C} = 37\text{ cal}$ or 155 J. Use the approximation that 1 jelly donut = 1 MJ.

(a) The ice water needed to offset 2 jelly donuts is

mass of water $= 2\text{ MJ}/(155\text{ J/g}) \approx \boxed{13\text{ kg}}$ of water or 13 liters.

(b) Assume that the body remains at $37°\text{C} = 310\text{ K}$.

$\Delta S_{water} = Q/T_{ave} = 2\text{ MJ}/292\text{ K} = 6.85 \times 10^3\text{ J/K}$

$\Delta S_{body} = Q/T = -2\text{ MJ}/310\text{ K} = -6.45 \times 10^3\text{ J/K}$.

$\Delta S_{total} = \boxed{400\text{ J/K}}$.

13.51 $$\Delta S = \frac{Q}{T} = \frac{mgh}{T} = \frac{1.0\text{ kg}(9.81\text{ m/s}^2)0.45\text{ m}}{300\text{ K}} = \boxed{15\text{ mJ/K}}.$$

13.53 $$\frac{Q_H}{Q_C} = \frac{T_H}{T_C}.\quad \text{So } Q_H = 80\text{ J}\,\frac{598\text{ K}}{393\text{ K}} = 122\text{ J}.$$

$W = Q_H - Q_C = 122\text{ J} - 80\text{ J} = \boxed{42\text{ J}}$.

13.55 (a) The efficiency is $\dfrac{W}{Q_H}$.

$$Q_H/t = \frac{W/t}{\text{efficiency}} = \frac{1000\text{ MW}}{0.333} = \boxed{3000\text{ MW}}.$$

(b) From conservation of energy: $Q_C/t = Q_H/t - W/t = \boxed{2000\text{ MW}}$.

(c) $P = Q_C/t = (m/t)c\Delta T$.

$\Delta T = 2000\text{ MW}/(6 \times 10^4\text{ kg/s} \times 4190\text{ J/kg°C}) = \boxed{8°\text{C}}$.

13.57 $Q_{C1} = Q_{H1} - W$ and efficiency is $e_1 = W/Q_{H1}$, so

$$Q_{C1} = W/e_1 - W = W\left(\frac{1}{e_1} - 1\right) \text{ and } Q_{C2} = W\left(\frac{1}{e_2} - 1\right).$$

$$\frac{Q_{C1}}{Q_{C2}} = \left(\frac{1 - e_1}{1 - e_2}\right)\left(\frac{e_2}{e_1}\right) = \frac{0.70}{0.60}\,\frac{0.40}{0.30} = 1.56.$$ $\boxed{\text{56\% more heat is released.}}$

13.59 (a) $\Delta S = Q/T = \sum \frac{mc\Delta T}{T}$. Compute this sum over the 2° intervals from 20°C

= 293 K to 40°C = 313 K. For the first 2 K interval,

$mc\Delta T$ = 1.0 kg (4190 J/kg·°C)(2°C) = 8380 J.

The average temperature is 294 K, so the first term in the sum is $\Delta S_1 = \frac{8380\text{ J}}{294\text{ K}}$.

When the similar terms are computed and summed we get

$$\Delta S = 8380\text{ J/K}\left(\frac{1}{294}+\frac{1}{296}+\frac{1}{298}+\frac{1}{300}+\frac{1}{302}\right)+$$

$$8380\text{ J/K}\left(\frac{1}{304}+\frac{1}{306}+\frac{1}{308}+\frac{1}{310}+\frac{1}{312}\right),$$

$\Delta S = \boxed{277\text{ J/K}}$.

(b) $\Delta S = Q/T$ = 1.0 kg(4190 J/kg°C)20°C/303 K = $\boxed{277\text{ J/K}}$.

(c) To three significant figures they are the same.

13.61 (a) From Chapter 11 we have the equation for heat flow:

$$\frac{\Delta Q}{\Delta t} = KA\frac{T_H - T_C}{L} = (398\text{ W/m·K})(4.0 \times 10^{-4}\text{ m})\frac{100\text{ K}}{0.30\text{ m}}$$

$$\frac{\Delta Q}{\Delta t} = \boxed{53\text{ W}}.$$

(b) $\frac{\Delta S}{\Delta t} = \frac{\Delta Q/\Delta t}{T} = \frac{-53\text{ W}}{373\text{ K}} = \boxed{-0.14\text{ W/K}}$.

(c) In the steady state condition the heat into the rod is equal to the heat out of the rod so that $\frac{\Delta S}{\Delta t} = \frac{\Delta Q/\Delta t}{T} = \frac{0}{T} = \boxed{0}$.

(d) $\frac{\Delta S}{\Delta t} = \frac{\Delta Q/\Delta t}{T} = \frac{53\text{ W}}{273\text{ K}} = \boxed{0.19\text{ W/K}}$

(e) $\frac{\Delta S}{\Delta t}$ = - 0.14 W/K + 0.19 W/K = $\boxed{0.05\text{ W/K}}$.

Chapter 14

PERIODIC MOTION

14.1 From Hooke's law: $F = kx$, so $x = F/k$. If the weight is evenly distributed, then each spring supports 1/4 of the total weight, then $F = \frac{15000\ N}{4} = 3750\ N$.

$$x = \frac{3750\ N}{7.00 \times 10^4\ N/m} = 5.35 \times 10^{-2}\ m = \boxed{5.35\ cm}.$$

14.3 From Hooke's law, the extension x due to the weight is
$x = F/k = (1.0\ kg)(9.8\ m/s^2)/(45\ N/m) = 0.22\ m = 22\ cm$.
The unloaded equilibrium length is 35 cm – 22 cm = $\boxed{13\ cm}$.

14.5 First find the equilibrium length with no load.
$x = F/k = (6.12\ kg)(9.81\ m/s^2)/(1000\ N/m) = 0.060\ m$.
Equilibrium length with no load is 0.18 m – 0.060 m = 0.12 m.
Find new x for 11.4 kg fish.
$x = F/k = (11.4\ kg)(9.81\ m/s^2)/(1000\ N/m) = 0.112\ m$.
Final length = 0.112 m + 0.120 m $\approx \boxed{23\ cm}$.

14.7 Because they have the same spring constant, the two springs will each stretch by the same amount x. Total length = initial lengths + stretch amounts.
1.30 m = 0.40 m + 0.25 m + 2x. $x = (0.65\ m)/2 = 0.325\ m$.
The connection point will be (0.25 + 0.325) m = $\boxed{0.575\ m}$ from the supported end of the shorter spring.

14.9 First get the value of $k = F/x_1 = mg/x_1$.
Next find the acceleration from $a = \frac{F}{m}$, where $F = kx$.

$$a = \frac{(mg/x_1)x}{m} = \frac{gx}{x_1} = \frac{9.81\ m/s^2 \times 0.0300\ m}{0.0750\ m} = \boxed{3.92\ m/s^2}.$$

14.11 The oscillator is described by a cosine function, $x = x_o \cos 2\pi t/T$.
$\cos 2\pi t/T = x/x_o = 27/30 = 0.90$,
$\cos^{-1} (0.90) = 0.451$ radians $= 2\pi t/T$.
$T = 2\pi(0.20\ s)/0.451 = \boxed{2.8\ s}$.

14.13 (a) If x = 0 at t = 0, then the displacement has the form of a sin function.
$x = x_o \sin 2\pi t/T = 7.0\ cm \sin (2\pi\ 0.063/0.314) = 7.0\ cm \times 0.95 = \boxed{6.7\ cm}$.
(b) Find x for t_2 and for t_1.
$x_2 = x_o \sin (2\pi\ 0.345/0.314) = 7.0\ cm \times 0.581 = 4.07\ cm$.
$x_1 = x_o \sin (2\pi\ 0.283/0.314) = 7.0\ cm \times (-0.581) = -4.07\ cm$.

The oscillator moves from -4.07 cm to +4.07 cm. The total distance traveled is 4.07 cm - (- 4.07 cm) = 8.14 cm ≈ 8.1 cm.

14.15 From the statement of the problem, T = 0.75 s, x_o = (0.75 m)/4.
$x = x_o \cos 2\pi t/T$. At t = 0, $x = x_o$. At t = 0.12 s,
$x = x_o \cos 2\pi[(0.12\text{ s})/(0.75\text{ s})] = 0.100$ m. The distance from the release point is $x_o - x = 0.75/4 - 0.100$ m = 0.087 m.

14.17 $W = PE = \frac{1}{2}kx^2$. $k = \dfrac{2\,PE}{x^2} = \dfrac{(2)\,(2\text{ J})}{(0.25\text{ m})^2} = 64\text{ N/m}$.

14.19 The spring in Ex.14.2 has a spring constant k = 9.48 N/m.
$W = \frac{1}{2}kx^2 = 0.5\,(9.48\text{ N/m})(0.050\text{ m})^2 = 1.2 \times 10^{-2}\text{ J}$.

14.21 (a) We have seen that $v_oT = 2\pi x_o$. So
$v_o = x_o\,2\pi/T = x_o 2\pi f = 10.0\text{ cm}(2\pi)(0.50/\text{s}) = 31\text{ cm/s}$.
(b) The maximum acceleration is given by $a_o = kx_o/m$. But conservation of energy shows that $\frac{1}{2}mv_o^2 = \frac{1}{2}kx_o^2$, so that $k/m = v_o^2/x_o^2$. Inserting this value for k/m in the expression for a_o gives
$a_o = v_o^2/x_o = (31\text{ cm/s})^2/10.0\text{ cm} = 99\text{ cm/s}^2$.

14.23 (a) Combine Newton's second law with Hooke's law to get $a = \dfrac{F}{m} = \dfrac{kx}{m}$.

$$k = \frac{ma}{x} = \frac{0.020\text{ kg} \times 9.8\text{ m/s}^2}{0.15\text{ m}} = 1.3\text{ N/m}.$$

(b) Use conservation of energy.

$$\frac{1}{2}mv_o^2 = \frac{1}{2}kx_o^2.\ \text{So } v_o = \sqrt{\frac{k}{m}}\,x_o = \sqrt{\frac{1.3\text{ N/m}}{0.020\text{ kg}}}\,0.15\text{ m} = 1.2\text{ m/s}.$$

14.25 (a) Inverting Eq.(14.8) we get

$$f = \frac{1}{T} = \frac{1}{2\pi}\sqrt{\frac{k}{m}} = \frac{1}{2\pi}\sqrt{\frac{2.00\text{ N/m}}{0.500\text{ kg}}} = \frac{1}{\pi} = 0.318\text{ Hz}.$$

(b) T = 1/f = 3.14 s.

14.27 (a) $\frac{1}{2}kx_o^2 = \frac{1}{2}mv_o^2$, so $v_o = x_o\sqrt{\dfrac{k}{m}} = 8.0\text{ cm}\sqrt{\dfrac{50\text{ N/m}}{2.0\text{ kg}}} = 40\text{ cm/s}$.

(b) $f = \dfrac{1}{T} = \dfrac{1}{2\pi}\sqrt{\dfrac{k}{m}} = \dfrac{5/\text{s}}{2\pi} = 0.80\text{ Hz}$.

14.29 (a) The amplitude is one half the total excustion, so $x_o = 0.50$ mm. The maximum speed is

$v_o = 2\pi x_o/T = 2\pi x_o f = 2\pi(0.50 \text{ mm})(440/\text{s}) = 1380 \text{ mm/s} = \boxed{1.38 \text{ m/s}}$.

(b) Maximum acceleration is $a_o = v_o^2/x_o = (1.38 \text{ m/s})^2/0.50 \text{ mm} = \boxed{3.8 \text{ km/s}^2}$.

14.31 The total energy is the maximum potential energy.

$E = \frac{1}{2} k x_o^2 = \frac{1}{2}(20 \text{ N/m})(0.050 \text{ m})^2 = 0.025$ J. The period of oscillation is

$$T = 2\pi\sqrt{\frac{m}{f}} = 2\pi\sqrt{\frac{0.10 \text{ kg}}{20 \text{ N/m}}} = 0.444 \text{ s}.$$

$x = x_o \cos(2\pi t/T) = x_o \cos(14.1\, t)$, where the angle is in radians.

$PE = \frac{1}{2} kx^2 = \frac{1}{2} k\left(x_o \cos(14.1\, t)\right)^2 = 0.025 \text{ J} \cos^2(14.1\, t)$.

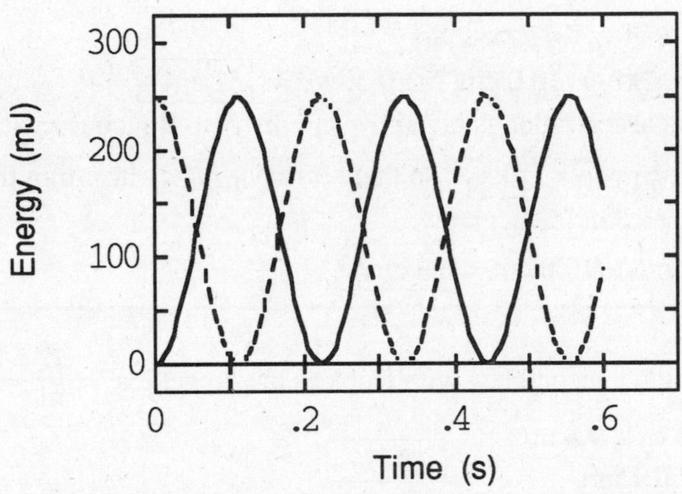

The dashed line is the PE, the solid line is the KE, and their sum is constant.

14.33 (a) Frequency is the number of oscillations divided by the time:

$$f = \frac{93}{60 \text{ s}} = \boxed{1.55 \text{ Hz}}.$$

(b) $T = 1/f = \boxed{0.645 \text{ s}}$.

14.35 (a) Period is $T = 2\pi\sqrt{L/g}$.

$$L = g\left(\frac{T}{2\pi}\right)^2 = 9.81 \text{ m/s}^2\left(\frac{0.750 \text{ s}}{2\pi}\right)^2 = \boxed{0.140 \text{ m}}.$$

(b) $L = 9.81 \text{ m/s}^2\left(\frac{2.00 \text{ s}}{2\pi}\right)^2 = \boxed{0.994 \text{ m}}$.

14.37 (a) Period is $T = 2\pi\sqrt{\frac{L}{g}} = 2\pi\sqrt{\frac{4.0\ \text{m}}{9.8\ \text{m/s}^2}} = \boxed{4.0\ \text{s}}$.

(b) Period is $T = 2\pi\sqrt{\frac{L}{g}} = 2\pi\sqrt{\frac{3.0\ \text{m}}{9.8\ \text{m/s}^2}} = \boxed{3.5\ \text{s}}$.

14.39 The period is $T = 2\pi\sqrt{\frac{L}{g}}$. Rearrange to get g. $\left(\frac{T}{2\pi}\right)^2 = \frac{L}{g}$,

$g = L\left(\frac{2\pi}{T}\right)^2 = 1.518\ \text{m}\left(\frac{2\pi}{2.475\ \text{s}}\right)^2 = \boxed{9.783\ \text{m/s}^2}$.

14.41 First find the period of the Tarzan pendulum.

$T = 2\pi\sqrt{\frac{L}{g}} = 2\pi\sqrt{\frac{21\ \text{m}}{9.8\ \text{m/s}^2}} = 9.2\ \text{s}.$

The time to swing across from one side to the other is T/2 = 4.6 s.

(a) Tarzan saves Jane 2.3 s before the bomb goes off.

(b) Since 2.3 s is 1/4 of the period, they will be over the middle of the river when the bomb goes off.

14.43 Start with the equation for the period, $T = 2\pi\sqrt{\frac{L}{g}}$, and note that f = 1/T.

$\frac{1}{f} = 2\pi\sqrt{\frac{L}{g}}$, so $\left(\frac{1}{2\pi f}\right)^2 = \frac{L}{g}$.

$L = g\left(\frac{1}{2\pi f}\right)^2 = \frac{981\ \text{cm/s}^2}{2.54\ \text{cm/in}}\left(\frac{1}{2\pi f\ \text{s}^{-1}}\right)^2$, where f is the frequency.

If we give the frequency in per minute V(per min) = 60 f (per sec), then f = V/60.

$L = 386.2\ \text{in}\left(\frac{60}{2\pi\ V}\right)^2 = \left(\frac{187.6}{V}\right)^2$ in.

14.45 $y = y_o e^{-\gamma t}\cos(2\pi t/T)$ For small angles the displacement $y = L\theta$.

First calculate the period $T = 2\pi\sqrt{\frac{L}{g}} = 2\pi\sqrt{\frac{1.26\ \text{m}}{9.81\ \text{m/s}^2}} = 2.25\ \text{s}.$

$\theta = \theta_o\, e^{-0.020 \times 1.50}\cos(2\pi\ 1.50/2.25) = \theta_o(0.97)(-0.50) = -0.49\ \theta_o.$

$\theta_o = 3.30°/-0.49 = \boxed{-6.76°}$.

14.47 Results depend on your particular situation.

14.49 You want the natural frequency of the pendulum to equal the resonant frequency of the spring–mass system.

$$f = \frac{1}{2\pi}\sqrt{\frac{g}{L}} = \frac{1}{2\pi}\sqrt{\frac{k}{m}}.$$

$L = gm/k = (9.8\ \text{m/s}^2)(1.0\ \text{kg})/(120\ \text{N/m}) = 0.082\ \text{m} = \boxed{8.2\ \text{cm}}$.

14.51 (a) $T = 2\pi\sqrt{m/k}$. $k = 4\pi^2 m/T^2 = 4\pi^2(0.500\ \text{kg})/(2.0\ \text{s})^2 = \boxed{4.9\ \text{N/m}}$.

(b) $F = kx = (4.9\ \text{N/m})(0.020\ \text{m}) = \boxed{0.098\ \text{N}}$.

14.53 The lower spring will stretch by 4.0 cm. Each upper spring supports only half the weight and so stretches by only 2.0 cm. Total extension is $\boxed{6.0\ \text{cm}}$.

14.55 $f = \frac{1}{2\pi}\sqrt{\frac{k}{m}}$ where $F = kx = mg$, so $k/m = g/x$.

$$f = \frac{1}{2\pi}\sqrt{\frac{9.8\ \text{m/s}^2}{0.067\ \text{m}}} = \boxed{1.9\ \text{Hz}}.$$

14.57 $x = x_o \cos 2\pi t/T$ and $v = -v_o \sin 2\pi t/T$.

$\cos^2 2\pi t/T + \sin^2 2\pi t/T = 1$. $(x/x_o)^2 + (v/v_o)^2 = 1$, so $(x/x_o)^2 = 1 - (v/v_o)^2$,

$\boxed{x = x_o\sqrt{1 - (v/v_o)^2}}$.

14.59 The gravitational force on a mass m at radial distance r from the center of a uniform earth is

$F = -G\frac{Mm}{r^2}$, where $M = \frac{4}{3}\pi\rho r^3$, and ρ is the density of the earth.

$\boxed{F = -\frac{4}{3}G\pi\rho m r = -kr}$, where $k = \frac{4}{3}G\pi\rho m$.

14.61 The period is $T = 2\pi\sqrt{\frac{L}{g}}$. The value of g is $\frac{GM}{R^2}$.

The ratio of the periods becomes

$$\frac{T_{moon}}{T_{earth}} = \sqrt{\frac{g_E}{g_m}} = \sqrt{\frac{M_E R_m^2}{M_m R_E^2}} = \frac{R_m}{R_E}\sqrt{\frac{M_E}{M_m}}$$

$$\frac{T_m}{T_E} = \frac{1.74 \times 10^6\ \text{m}}{6.38 \times 10^6\ \text{m}}\sqrt{\frac{5.98 \times 10^{24}\ \text{kg}}{7.36 \times 10^{22}\ \text{kg}}} = 2.46$$

$T_m = 2.46\ T_E = 2.46\ (2.00\ \text{s}) = \boxed{4.92\ \text{s}}$.

14.63 (a) Let the water be displaced so that one column is a height x above the other.

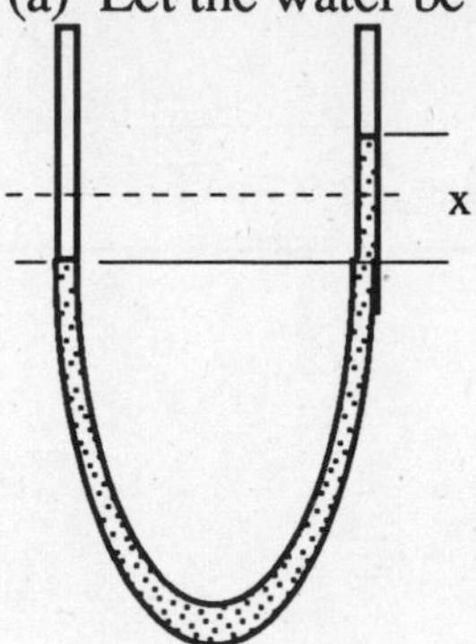

The restoring force is $F = -\rho g V = -\rho g \frac{\pi d^2}{4} x$, where d is the inner diameter of the tube assumed constant throughout. This force acts on the total mass $m = \rho \frac{\pi d^2}{4} L$. From Newton's second law we get

$$a = \frac{-\rho g \frac{\pi d^2}{4} x}{\rho \frac{\pi d^2}{4} L} = -\frac{g}{L} x.$$ This is the harmonic oscillator equation.

(b) The frequency of the oscillation is $f = \frac{1}{2\pi}\sqrt{\frac{g}{L}}$.

The period is $T = \frac{1}{f} = \boxed{2\pi\sqrt{\frac{L}{g}}}$.

14.65 Bobbing frequency: $f_b = \frac{1}{2\pi}\sqrt{\frac{k}{m}}$, where $F = mg = kx$, so $\frac{k}{m} = \frac{g}{0.02} = 50\text{ g}$.

So $f_b = \frac{1}{2\pi}\sqrt{50\text{ g}}$.

Swing frequency: $f_s = \frac{1}{2\pi}\sqrt{\frac{g}{L}} = \frac{f_b}{10}$.

$\sqrt{\frac{g}{L}} = \frac{\sqrt{50\text{ g}}}{10}$. Square both sides to get $\frac{1}{L} = \frac{50}{100}$. $L = \boxed{2.0\text{ m}}$.

14.67 The period of a pendulum is given by Eq.(14.11) as $\tau = 2\pi\sqrt{\frac{L}{g}}$.
At initial temperature T the length of the wire is L_o. At temperature $T + \Delta T$ the length of the pendulum becomes $L = L_o(1 + \alpha\Delta T)$. So the initial period is

$\tau_o = 2\pi\sqrt{\frac{L_o}{g}}$ and the new period is $\tau = 2\pi\sqrt{\frac{L_o(1+\alpha\Delta T)}{g}}$.

$$\Delta\tau = \tau - \tau_o = 2\pi\sqrt{\frac{L_o}{g}}\left(\sqrt{1+\alpha\Delta T} - 1\right),$$

$$\Delta\tau = \tau_o\left((1+\alpha\Delta T)^{1/2} - 1\right) \approx \tau_o\left(1 + \tfrac{1}{2}\alpha\Delta T - 1\right) = \boxed{\tfrac{1}{2}\alpha\tau_o\Delta T}.$$

Chapter 15

WAVE MOTION

15.1 We are given that $\lambda = 1.5$ m and v = 10 m/s. Use Eq.(15.2) to find the frequency.

$$f = \frac{v}{\lambda} = \frac{10 \text{ m/s}}{1.5 \text{ m}} = \boxed{6.7 \text{ Hz}}.$$

15.3 Start with Eq.(15.1),

$y(x,t) = y_o \sin 2\pi\left(\frac{x}{\lambda} - \frac{t}{T}\right)$, where $y_o = 3.0$ cm and $\lambda = 40$ cm. The period is found from $T = 1/f = \lambda/v = (40 \text{ cm})/(40 \text{ cm/s}) = 1.0$ s.

(a) $y(0.0 \text{ cm}, 2.0 \text{ s}) = 3.0 \text{ cm} \sin 2\pi(0 - 2) = \boxed{0}$.

(b) $y(10 \text{ cm}, 20 \text{ s}) = 3.0 \text{ cm} \sin 2\pi\left(\frac{10}{40} - \frac{20}{1.0}\right) = 3.0 \text{ cm} \sin\frac{\pi}{2} = \boxed{3.0 \text{ cm}}$.

15.5 Use Eq.(15.2) to find the frequency.

$$f_1 = \frac{c}{\lambda} = \frac{3.00 \times 10^8 \text{ m/s}}{90 \text{ m}} = 3.3 \text{ MHz}.$$

$$f_2 = \frac{c}{\lambda} = \frac{3.00 \times 10^8 \text{ m/s}}{11 \text{ m}} = 27 \text{ MHz}.$$

Frequency range is $\boxed{3.3 \text{ MHz} - 27 \text{ MHz}}$.

15.7 Start with Eq.(15.1). $y(x,t) = y_o \sin 2\pi\left(\frac{x}{\lambda} - \frac{t}{T}\right)$

$y_o = 10.0$ cm and $\lambda = 8.00$ cm. When x = 15.0 cm and t = 2.00 s, then y = 8.66 cm. Use this to find T.

$$y(15.0 \text{ cm}, 2.00 \text{ s}) = 8.66 \text{ cm} = 10.0 \text{ cm} \sin 2\pi\left(\frac{15.0}{8.00} - \frac{2.00}{T}\right),$$

$$2\pi\left(\frac{15.0}{8.00} - \frac{2.00}{T}\right) = \sin^{-1}(0.866),$$

$$2\pi\left(\frac{15.0}{8.00} - \frac{2.00}{T}\right) = 1.05 \text{ rad}.$$

$$\frac{2.00}{T} = \frac{15.0}{8.00} - \frac{1.05}{2\pi} = 1.71. \qquad T = \boxed{1.17 \text{ s}}.$$

15.9 The amplitude of the wave is $y_o = 6.0$ cm. We compute the period from

$T = 2\pi\sqrt{\frac{m}{k}} = 2\pi\sqrt{\frac{2.3 \text{ kg}}{20 \text{ N/m}}} = 2.13$ s. From this we can get the wavelength.

$\lambda = vT = (0.30 \text{ m/s})(2.13 \text{ s}) = 0.64$ m.

At t = 0, x = 0, and $y = -y_o$. The resulting wave will be a cosine wave with a negative initial displacement.

$$y(x,t) = -y_o \cos 2\pi\left(\frac{x}{0.64\ \text{m}} - \frac{t}{2.1\ \text{s}}\right) = \boxed{y_o \sin\left(\frac{2\pi x}{0.64\ \text{m}} - \frac{2\pi t}{2.1\ \text{s}} - \frac{\pi}{2}\right)}.$$

15.11 The power transmitted depends on the amplitude and the frequency, $P \propto y_o^2 f^2$.

For constant power, $y_{o1}^2 f_1^2 = y_{o2}^2 f_2^2$.

$$f_2 = f_1 \frac{y_{o1}}{y_{o2}} = f_1 \frac{y_{o1}}{y_{o1}/3} = 3f_1.$$

$\boxed{\text{The child must triple the frequency of the waves.}}$

15.13 The equation for the power transmitted is $P = 2\pi^2 A\rho v f^2 y_o^2$.

Here ρ = mass/volume and A is the area, so ρA = mass/length = μ.

$2\pi f = \omega$. These substitutions yield $P = \frac{1}{2}(2\pi f)^2 \mu v y_o^2 = \boxed{\frac{1}{2}\mu v \omega^2 y_o^2}$.

15.15 Use Eq.(15.2), $f = v/\lambda$.

(a) For $\lambda = 3.4$ cm, $f = (340\ \text{m/s})/0.034\ \text{m} = \boxed{10\ 000\ \text{Hz}}$.

For $\lambda = 34$ cm, $f = (340\ \text{m/s})/0.34\ \text{m} = \boxed{1000\ \text{Hz}}$.

For $\lambda = 3.4$ cm, $f = (340\ \text{m/s})/3.4\ \text{m} = \boxed{100\ \text{Hz}}$.

For $\lambda = 3.4$ cm, $f = (340\ \text{m/s})/34\ \text{m} = \boxed{10\ \text{Hz}}$.

(b) $\boxed{\text{All of these can be heard except the 10 Hz frequency.}}$

15.17 The expression for the speed of sound was given in the text as $v(T) = (331.5 + 0.6T)$ m/s, where T is in °C. At T = 35°C we get $v = (331.5 + 0.6 \times 35)\ \text{m/s} = \boxed{352.5\ \text{m/s}}$.

15.19 From Eq.(15.4): $\frac{I_1}{I_2} = \frac{6}{1} = \frac{r_2^2}{r_1^2} = \frac{r_2^2}{(8.0\ \text{m})^2}$,

$$r_2 = 8.0\ \text{m}\sqrt{6} = 19.6\ \text{m} \approx \boxed{20\ \text{m}}.$$

15.21 $I \propto y^2$ and $I \propto 1/r^2$.

$\frac{I_1}{I_2} = \frac{y_1^2}{y_2^2} = \frac{r_2^2}{r_1^2}$. Taking square roots we get $\boxed{\frac{y_1}{y_2} = \frac{r_2}{r_1}}$.

15.23 The expression for the speed of sound was given in the text as $v = (331.5 + 0.6T)$ m/s when T is in °C. The relationship between Celsius temperature and Fahrenheit temperature is $T(°C) = \frac{5}{9}[T(°F) - 32]$.

$$v = \left(331.5 + 0.6 \times \frac{5}{9}[T - 19.2]\right) \text{m/s (3.28 ft/m)},$$

$$v = \boxed{(1066 + 1.1\,T(°F))\ \text{ft/s}}.$$

15.25 In dB, gain = $10 \log(I/I_o) = 10 \log(100\,000) = \boxed{50 \text{ dB}}$.

15.27 Find the ratio of I/I_o.
$65 \text{ dB} = 10 \log(I/I_o)$,
$I/I_o = 10^{6.5} = \boxed{3.16 \times 10^6}$.

15.29 $68 \text{ dB} = 10 \log(I/I_o)$,
$I/I_o = 10^{6.8}$.
$I = I_o \times 10^{6.8} = 10^{-12} \text{ W/m}^2 \times 6.31 \times 10^6 = \boxed{6.3 \times 10^{-6} \text{ W/m}^2}$.

15.31 Assume the intensity of one loom is I. Then $85 \text{ dB} = 10 \log 5I/I_o$. Solve for I/I_o.
$I/I_o = (10^{8.5})/5 = 6.325 \times 10^7$.
$10 \log 7I/I_o = \boxed{86.5 \text{ dB}}$.

15.33 (a) Use the doppler formula for an approaching source.

$$f' = \frac{f}{1 - v_s/v} = \frac{2000 \text{ Hz}}{1 - 40/340} = \boxed{2270 \text{ Hz}}.$$

(b) $$f'_{(leaving)} = \frac{f}{1 + v_s/v} = \frac{2000 \text{ Hz}}{1 + 40/340} = 1790 \text{ Hz}.$$

$\Delta f = 2270 - 1790 \text{ Hz} = \boxed{480 \text{ Hz}}$.

15.35 First note that $80 \text{ km/h} = 80 \text{ km/h} \frac{1000 \text{ m/km}}{3600 \text{ s/h}} = 22.2 \text{ m/s}$.

Approaching: $$f' = \frac{f}{1 - v_s/v} = \frac{750 \text{ Hz}}{1 - 22.2/340} = 802 \text{ Hz}.$$

Leaving: $$f' = \frac{f}{1 + v_s/v} = \frac{750 \text{ Hz}}{1 + 22.2/340} = 704 \text{ Hz}.$$

$\Delta f' = 802 \text{ Hz} - 704 \text{ Hz} = \boxed{98 \text{ Hz}}$.

15.37 Use the Doppler formula for combined motion of both source and observer.

$$f' = f\left(\frac{1 - v_o/v}{1 - v_s/v}\right) = 1500 \text{ Hz}\left(\frac{1 - 25/340}{1 - 30/340}\right) = \boxed{1524 \text{ Hz}}.$$

15.39 First find v_s. $v_s = 2\pi r/T = 2\pi(0.75 \text{ m})/(1/3 \text{ s}) = 14 \text{ m/s}$. At the extremes of approaching and departing,

$f' = f/(1 \pm v_s/v) = f/(1 \pm 0.041) \approx f(1 \pm 0.04)$.

As a function of time the frequency is approximately

$f' = f(1 + 0.04 \cos 2\pi t/T) = \boxed{800 \text{ Hz}(1 + 0.04 \cos 6\pi t)}$.

15.41 From Eq.(15.8) we get

$v = v_s \sin\theta = (400 \text{ m/s}) \sin 65° = \boxed{363 \text{ m/s}}$.

15.43 From Eq.(15.8), $v/v_s = \sin 37°$. $v_s/v = 1/\sin 37° = 1/0.602 = 1.66$.

The speed is $\boxed{\text{Mach } 1.66}$.

15.45 First find v_s from Eq.(15.8). $v_s = v/\sin\theta = 961/\sin 45.0° = 1359$ m/s.

In air, $\sin\theta = (340 \text{ m/s})/(1359 \text{ m/s}) = 0.250$,

$\theta = \sin^{-1}(0.250) = \boxed{14.5°}$.

15.47 The fundamental wavelength is $\lambda = 2L = 2(0.65 \text{ m}) = 1.30$ m.

$v = f\lambda = (440 \text{ Hz})(1.30 \text{ m}) = \boxed{572 \text{ m/s}}$.

15.49 For the longer string: $f_1 = v/2L$.

For the shorter string, second harmonic is: $f_2 = 2v/2L_2 = v/(L/1.44)$.

$$\frac{f_1}{f_2} = \frac{v/2L}{1.44v/L} = \frac{1}{2 \times 1.44} = \frac{1}{2.88} = \boxed{0.35}.$$

15.51 (a) $v = \sqrt{\frac{T}{\mu}} = \sqrt{\frac{850 \text{ N}}{0.030 \text{ kg}/1.50 \text{ m}}} = \boxed{206 \text{ m/s}}$.

(b) $\frac{v}{v_{air}} = \frac{206 \text{ m/s}}{340 \text{ m/s}} = \boxed{0.61}$.

15.53 Four antinodes implies four half–wavelengths or 2 wavelengths $= 2\lambda = L$.

$\lambda = L/2 = v/f$.

$L = 2v/f = 2(400 \text{ m/s})/800 \text{ Hz} = \boxed{1.00 \text{ m}}$.

15.55 Each semitone has a frequency $\sqrt[12]{2}$ times the frequency of the preceding tone.

A difference of 16 semitones is

$f/f_o = \left(\sqrt[12]{2}\right)^{16} = \boxed{2.52}$.

15.57 The tension in the string is $T = M(g + a)$. The frequency of the vibration is

$$f = \frac{v}{2L} = \frac{\sqrt{\frac{TL}{m}}}{2L} = \frac{1}{2}\sqrt{\frac{T}{mL}} = \frac{1}{2}\sqrt{\frac{M(g + a)}{mL}},$$

$$f = \frac{1}{2}\sqrt{\frac{M(1.10 \text{ g})}{mL}} = \frac{1}{2}\sqrt{\frac{(5.0 \text{ kg})(1.10 \times 9.81 \text{ m/s}^2)}{(0.010 \text{ kg})(1.00 \text{ m})}} = \boxed{36.7 \text{ Hz}}.$$

15.59 (a) $f_1 = v/\lambda = v/2L = (340 \text{ m/s})/2(0.50 \text{ m}) = \boxed{340 \text{ Hz}}$.

(b) With tube closed, $\lambda = 4L$, so $f_1 = (340 \text{ m/s})/2(1.00 \text{ m}) = \boxed{170 \text{ Hz}}$.

15.61 Fundamental frequency $= f_1 = v/4L = (340 \text{ m/z})/4(0.25 \text{ m}) = \boxed{340 \text{ Hz}}$.
Overtones present are the odd harmonics, 3, 5, 7.
The frequencies are $f_3 = \boxed{1020 \text{ Hz}}$, $f_5 = \boxed{1700 \text{ Hz}}$, and $f_7 = \boxed{2380 \text{ Hz}}$.

15.63 The beat frequency is $f_b = f_2 - f_1$. Let f_2 be the higher frequency.
Then $f_1 = f_2 - f_b = 262 - 2 = \boxed{260 \text{ Hz}}$.

15.65 Fundamental frequency is $f_{higher} - f_{beat} = 398$ Hz.
$f = v/2L$, so $L = v/2f = (150 \text{ m/s})/2(398 \text{ Hz}) = 0.1884$ m. For the shorter string, $L = (150 \text{ m/s})/2(400 \text{ Hz}) = 0.1875$ m. The difference is
$\Delta L = 0.1884 - 0.1875 = 0.0009 \text{ m} = \boxed{0.9 \text{ mm}}$.

15.67 First calculate L_o. $f_1 = v/2L$, so $L_o = v/2f$.
At 8°C, $v = 336.3$ m/s, and at 27°C, $v = 347.7$ Hz.
$L_o = (336.3 \text{ m/s})/(2 \times 180 \text{ /s}) = 0.9342$ m.
At the higher temperatures the lengths are
$L_{cu} = L_o(1 + \alpha\Delta T) = 0.934 \text{ m}(1 + 17 \times 10^{-6} \times 19) = 0.9342 \text{ m}(1.000323)$,
$L_{steel} = L_o(1 + \alpha\Delta T) = 0.934 \text{ m}(1 + 12 \times 10^{-6} \times 19) = 0.9342 \text{ m}(1.000228)$,
$f_{cu} = v/2L = 347.7/2(0.9342)(1.000323) = 186.035$ Hz,
$f_{steel} = v/2L = 347.7/2(0.9342)(1.000228) = 186.053$ Hz.
Beat frequency $= f_b = f_{steel} - f_{cu} = \boxed{0.018 \text{ Hz}}$.

15.69 Find the ratio of I/I_o.
$25 \text{ dB} = 10 \log(I/I_o)$,
$I/I_o = 10^{2.5} = \boxed{316}$.

15.71 (a) From Eq.(15.6) we get $f' = \dfrac{f}{1 - v_s/v} = \dfrac{1000 \text{ Hz}}{1 - 20/340} = \boxed{1062.5 \text{ Hz}}$.
(b) Calculate the frequency with the velocities relative to the air.
$f' = f(1 + v_o/v)/(1 - v_s/v) = 1000 \text{ Hz}(1 + 5/340)/(1 - 15/340) = \boxed{1062 \text{ Hz}}$.

15.73 (a) $v = \sqrt{\dfrac{T}{\mu}} = \lambda f$. Here $\mu = m/L$ and $\lambda = 2L$.
$T/\mu = \lambda^2 f^2$, so $T = (\mu)(2L)^2 f^2$, $T = 4\mu L^2 f^2$.
$T_1 = 4(5.05 \times 10^{-3} \text{ kg/m})(0.429 \text{ m}^2)(82/\text{s})^2 = \boxed{58 \text{ N}}$.
$T_2 = 4(3.71 \times 10^{-3} \text{ kg/m})(0.429 \text{ m}^2)(110/\text{s})^2 = \boxed{77 \text{ N}}$.
$T_3 = 4(2.21 \times 10^{-3} \text{ kg/m})(0.429 \text{ m}^2)(147/\text{s})^2 = \boxed{82 \text{ N}}$.
$T_4 = 4(1.01 \times 10^{-3} \text{ kg/m})(0.429 \text{ m}^2)(196/\text{s})^2 = \boxed{67 \text{ N}}$.

$T_5 = 4(0.58 \times 10^{-3}\ \text{kg/m})(0.429\ \text{m}^2)(247/\text{s})^2 = \boxed{61\ \text{N}}$.

$T_6 = 4(0.44 \times 10^{-3}\ \text{kg/m})(0.429\ \text{m}^2)(330/\text{s})^2 = \boxed{82\ \text{N}}$.

(b) Total force is the sum = $\boxed{427\ \text{N}}$.

15.75 Combine Eq.(15.6) and Eq.(15.7) to get

(a) $f' = f\left(\dfrac{1 + v_o/v}{1 - v_s/v}\right) = 1000\ \text{Hz}\left(\dfrac{1 + 40/340}{1 - 30/340}\right) = \boxed{1230\ \text{Hz}}$.

(b) $f' = f\left(\dfrac{1 - v_o/v}{1 - v_s/v}\right) = 1000\ \text{Hz}\left(\dfrac{1 - 40/340}{1 - 30/340}\right) = \boxed{968\ \text{Hz}}$.

15.77 $\lambda = v/f$. The change in λ is $\Delta\lambda = \lambda' - \lambda = (v'/v)\lambda - \lambda$,

$\Delta\lambda = \dfrac{v' - v}{v}\lambda \approx \dfrac{0.6\ \Delta T}{331.5}\lambda = \boxed{(1.8 \times 10^{-3})\ \lambda\ \Delta T}$.

15.79 $y_1 = y_o \sin 2\pi(x/\lambda - t/T)$ and $y_2 = y_o \sin 2\pi(x/\lambda + t/T)$.

$y = y_1 + y_2 = y_o\,[\sin 2\pi(x/\lambda - t/T) + \sin 2\pi(x/\lambda + t/T)]$

Use the relationship $\sin A + \sin B = 2\cos\left(\dfrac{A - B}{2}\right) \cdot \sin\left(\dfrac{A + B}{2}\right)$.

Let $A = 2\pi\left(\dfrac{x}{\lambda} - \dfrac{t}{T}\right)$ and $B = 2\pi\left(\dfrac{x}{\lambda} + \dfrac{t}{T}\right)$

Then $\dfrac{A - B}{2} = 2\pi\dfrac{x}{\lambda}$ and $\dfrac{A + B}{2} = 2\pi\dfrac{t}{T}$. So,

$$\boxed{y = 2y_o \cos\left(2\pi\frac{x}{\lambda}\right) \cdot \sin\left(2\pi\frac{t}{T}\right)}$$

Chapter 16

ELECTRIC CHARGE AND ELECTRIC FIELD

16.1 We can find the force from Coulomb's law, Eq. (16.1).

$$F = k\frac{q_1 q_2}{r^2} = 9.0 \times 10^9\ \text{N·m}^2/\text{C}^2\ \frac{(2.0 \times 10^{-6}\ \text{C})(-4.0 \times 10^{-6}\ \text{C})}{(0.15\ \text{m})^2},$$

$F = \boxed{-3.2\ \text{N}}$. The forces are $\boxed{\text{attractive}}$.

16.3 Use Coulomb's law. $F = k\frac{q^2}{x^2}$.

$$x = q\sqrt{\frac{k}{F}} = 2.6 \times 10^{-6}\ \text{C}\sqrt{\frac{9.0 \times 10^9\ \text{N·m}^2/\text{C}^2}{2.0 \times 10^{-8}\ \text{N}}} = \boxed{1.7\ \text{km}}.$$

16.5 Use Coulomb's law and find the ratio of the forces. Let subscript 1 correspond to the original situation and subscript 2 correspond to the new situation.

$$\frac{F_2}{F_1} = \left(\frac{q_2^2}{q_1^2}\right)\left(\frac{r_1^2}{r_2^2}\right).$$ The charge remains constant so $q_1 = q_2$. Since $\frac{F_2}{F_1} = 10$,

$$\frac{r_1^2}{r_2^2} = 10. \quad \frac{r_2}{r_1} = \sqrt{\frac{1}{10}} = \boxed{0.32}.$$

16.7 (a) Use Newton's law of gravitation and Coulomb's law of electricity.

$$\frac{F_{elec}}{F_{grav}} = \frac{kq^2/r^2}{Gm^2/r^2} = \frac{8.99 \times 10^9\ \text{N·m}^2/\text{C}^2}{6.67 \times 10^{-11}\ \text{N·m}^2/\text{kg}^2}\ \frac{(1.60 \times 10^{-19}\ \text{C})^2}{(1.67 \times 10^{-27}\ \text{kg})^2},$$

$$\frac{F_{elec}}{F_{grav}} = \boxed{1.24 \times 10^{36}}.$$

(b) $\boxed{\text{No.}}$ The gravitational force of attraction is far smaller than the repulsive electrostatic force.

16.9 $$F = k\frac{q_1 q_2}{r^2} = 8.99 \times 10^9\ \text{N·m}^2/\text{C}^2\ \frac{(1.60 \times 10^{-19}\ \text{C})(-1.60 \times 10^{-19}\ \text{C})}{(2.50 \times 10^{-10}\ \text{m})^2},$$

$F = \boxed{-3.68 \times 10^{-9}\ \text{N}}$. Note that the force is attractive because the charges have opposite signs.

16.11 Set the Coulomb force equal to the gravitational force: $k\frac{q^2}{r^2} = mg$.

$$r = q\sqrt{\frac{k}{mg}},$$

$$r = 1.60 \times 10^{-19}\ \text{C}\sqrt{\frac{8.99 \times 10^9\ \text{N·m}^2/\text{C}^2}{1.67 \times 10^{-27}\ \text{kg} \times 9.81\ \text{m/s}^2}} = \boxed{0.119\ \text{m}}.$$

16.13 The force on the spheres is given by Coulomb's law.

$F = k\frac{q_1 q_2}{r^2}$, and $q_1 + q_2 = 6.00\ \mu C = 6.00 \times 10^{-6}$ C.

Rearrange Coulombs law to get

$$q_1(6.00 \times 10^{-6}\ C - q_1) = \frac{Fr^2}{k},$$

$$q_1(6.00 \times 10^{-6}\ C - q_1) = \frac{(1.52\ N)(0.200\ m)^2}{8.99 \times 10^9\ N{\cdot}m^2/C^2}.$$

Rearrange to get a quadratic equation.

$$q_1^2 - 6.00 \times 10^{-6}\ q_1 + 6.76 \times 10^{-12} = 0.$$

Using the quadratic formula we find the value of q_1 to be

$$q_1 = \frac{6.00 \times 10^{-6} \pm \sqrt{36.0 \times 10^{-12} - 4(6.76 \times 10^{-12})}}{2}\ C,$$

$$q_1 = \frac{6.00 \times 10^{-6} C \pm 2.99 \times 10^{-6}\ C}{2} = 1.50\ \mu C \text{ or } 4.50\ \mu C.$$

Since these two values add to the initial 6.00 μC, then the two charges must be $\boxed{1.50\ \mu C \text{ and } 4.50\ \mu C}$.

16.15 First find the charge on each ball when they are 4.0 cm apart. Refer to the figure for the solution of Problem 16.14. The angle θ is given by

$$\theta = \sin^{-1}\frac{d/2}{L} = \sin^{-1}\frac{2.0}{30} = 3.8°.$$

$$\tan\theta = \frac{F_e}{F_g} = \frac{kq^2/d^2}{mg}.$$

$$q = d\sqrt{\frac{mg \tan 7.7°}{k}} = 4.0 \times 10^{-2}\ m\sqrt{\frac{1.0 \times 10^{-4}\ kg(9.8\ m/s^2)\tan 3.8°}{9.0 \times 10^9\ N{\cdot}m^2/C^2}}$$

$$q = 3.4 \times 10^{-9}\ C = 3.4\ nC.$$

The rate of charge leaking off is

$$\frac{\Delta q}{\Delta t} = \frac{8.0\ nC - 3.4\ nC}{2\ h} = \boxed{2.3\ nC/h}.$$

16.17 First make a sketch of the situation.

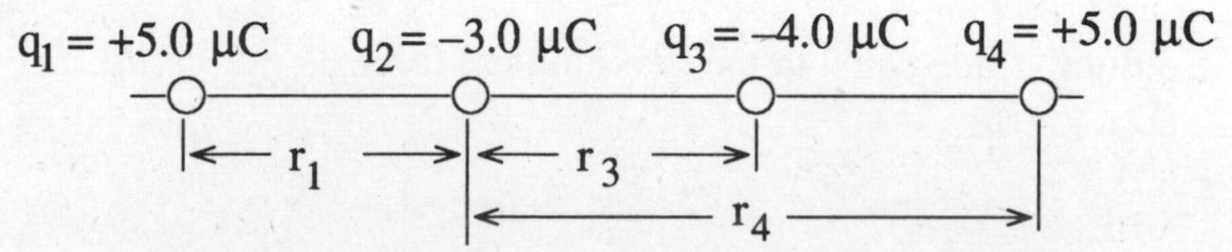

$q_1 = +5.0\ \mu C$ $q_2 = -3.0\ \mu C$ $q_3 = -4.0\ \mu C$ $q_4 = +5.0\ \mu C$

r_1 r_3

r_4

$$F = kq_2\left(\frac{q_1}{r_1^2} - \frac{q_3}{r_3^2} - \frac{q_4}{r_4^2}\right).$$

The signs shown in the equation above would give repulsive force to the right from q_1 if q_1 and q_2 were both positive. The forces from q_3 and q_4 would be to the left (negative) if all of the qs were positive. Thus a net positive force would be to the right and a net negative force would be to the left. Substituting in the numerical values gives

$$F = 9.0 \times 10^9\ \text{N}\cdot\text{m}^2/\text{C}^2(-3.0\ \mu\text{C})\left(\frac{+5.0}{(0.50)^2} - \frac{-4.0}{(0.50)^2} - \frac{+5.0}{(1.0)^2}\right)\frac{\mu\text{C}}{\text{m}^2},$$

$$F = 9.0 \times 10^9\ \text{N/C}^2(-3.0 \times 10^{-6}\ \text{C})(20 + 16 - 5)(10^{-6}\ \text{C}),$$

$$F = \boxed{-0.84\ \text{N (toward the origin)}}.$$

16.19 A sketch of the situation is shown below.

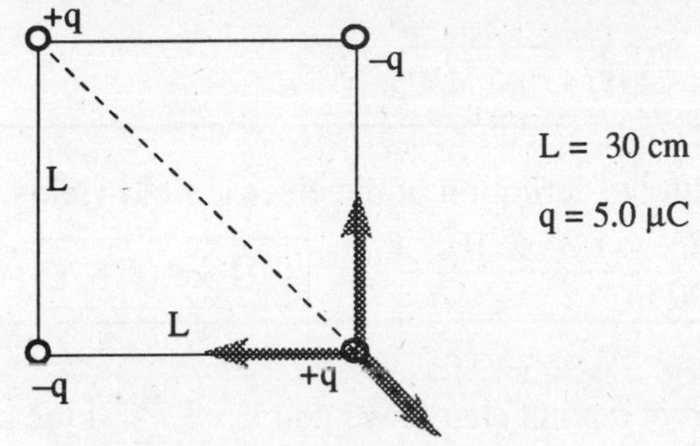

The magnitude of the force from the nearest charge (on the left) is $k\frac{q^2}{L^2}$ directed toward the left. Another force of the same magnitude is due to the charge above and this force is directed upward in the diagram. These forces add up to $F = \sqrt{2}\,k\frac{q^2}{L^2}$ along the diagonal toward the opposite corner. The charge diagonally opposite gives a force $k\frac{q^2}{2L^2}$, because the separation is $\sqrt{2}\,L$. This force is outward along the diagonal. The total force is $F_{net} = \left(\sqrt{2} - \frac{1}{2}\right)k\frac{q^2}{L^2}$ and is directed inward along the diagonal. When the numerical values are substituted we get

$$F_{net} = (0.91)(9.0 \times 10^9\ \text{N}\cdot\text{m}^2/\text{C}^2)\,\frac{(5.0 \times 10^{-6}\ \text{C})^2}{(0.30\ \text{m})^2},$$

$\boxed{F_{net} = 2.3\ \text{N diagonally inward.}}$

16.21 A sketch of the situation is shown below.

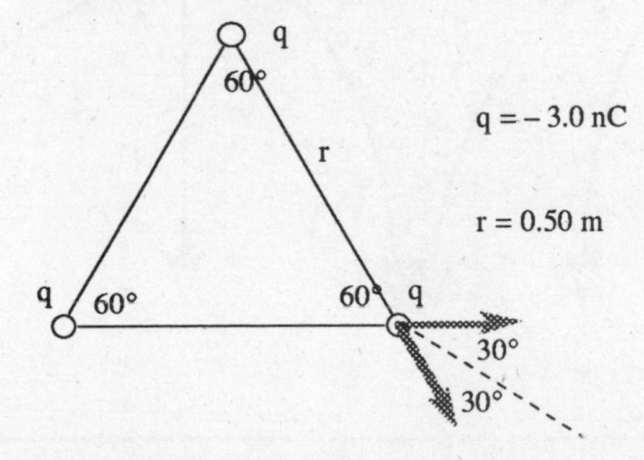

The force between each pair of charges is $F = k\frac{q^2}{r^2}$. From the symmetry of the diagram we see that the sum of the two forces will lie along the dashed line. The vector sum of the forces is $F_{net} = 2\ F \cos 30°$.

$$F_{net} = 2\ (9.0 \times 10^9\ N{\cdot}m^2/C^2) \left(\frac{-3.0 \times 10^{-9}\ C}{0.50\ m}\right)^2 \cos 30°,$$

$$F_{net} = \boxed{5.6 \times 10^{-7}\ N \text{ outward from the center of the triangle}}.$$

16.23 The field is defined by $E = F/q$.

$E = (1.00\ N)/1.25 \times 10^{-6}\ C) = \boxed{8.00 \times 10^5\ N/C}$.

16.25 Combining Coulomb's law with the definition of the electric field yields

$$E = \frac{kq}{r^2} = \frac{8.99 \times 10^9\ N{\cdot}m^2/C^2 \times 1.35 \times 10^{-6}\ C}{(0.100\ m)^2} = \boxed{1.21 \times 10^6\ N/C}.$$

16.27 From the definition of the field of a point charge we get $E = k\frac{q}{r^2}$. If the charge is negative then the field is negative indicating that the field is directed toward the source charge. Thus

$$q = -\frac{Er^2}{k} = -\frac{(90.5\ N/C)(1.56\ m)^2}{8.99 \times 10^9\ N{\cdot}m^2/C^2} = \boxed{-2.45 \times 10^{-8}\ C}.$$

16.29 A sketch of the electric field is shown below.

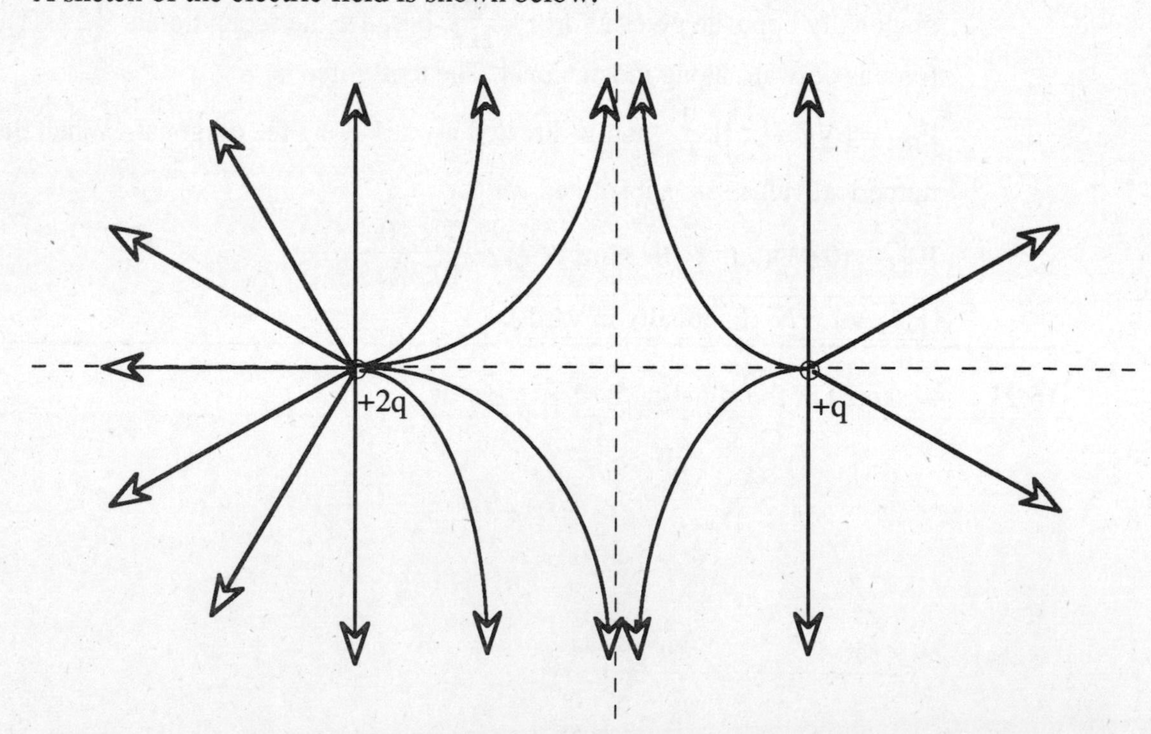

16.31 The angle of the field from the x axis is given by

$$\theta = \tan^{-1}\left(\frac{E_y}{E_x}\right) = \tan^{-1}\left(\frac{8}{6}\right) = 53.1°.$$

The magnitude of the field is

$$E = \sqrt{E_x{}^2 + E_y{}^2} = 10000 \text{ N/C}.$$

$$F = qE = \boxed{1.50 \times 10^{-2} \text{ N at } 53.1°}.$$

16.33 A sketch of the physical situation is shown below.

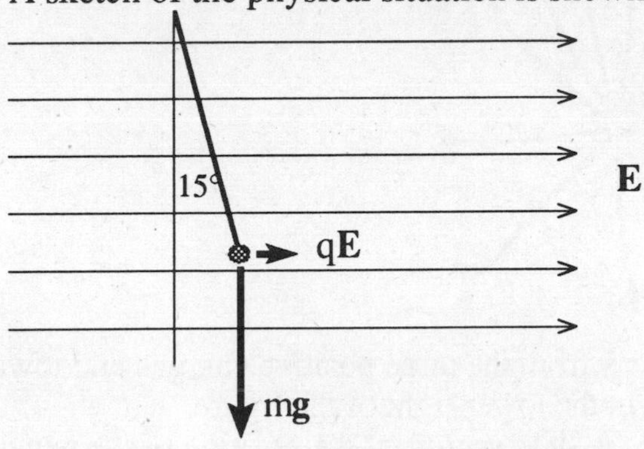

At equilibrium, the angle θ is found from: $\tan\theta = \frac{F_e}{F_g} = \frac{qE}{mg}$.

$$q = \frac{mg}{E}\tan\theta = \frac{(0.500 \times 10^{-3} \text{ kg})(9.81 \text{ m/s}^2)}{400 \text{ N/C}}\tan 15° = \boxed{3.29\ \mu\text{C}}.$$

16.35 (a) $F = qE = 1.60 \times 10^{-19} \text{ C} \times 5000 \text{ N/C} = \boxed{8.00 \times 10^{-16} \text{ N}}$.

(b) $KE = \frac{1}{2}mv^2 = \frac{1}{2}m(2ax) = m\frac{F}{m}x = Fx.$

$$KE = (8.00 \times 10^{-16} \text{ N})(4.00 \times 10^{-2} \text{ m}) = \boxed{3.2 \times 10^{-17} \text{ J}}.$$

16.37 Simply add the fields as vectors to get

$$E = \sqrt{E_N{}^2 + E_E{}^2} = \sqrt{(12.6)^2 + (17.4)^2} \text{ N/C} = 21.5 \text{ N/C}.$$

$$\theta = \tan^{-1}\frac{E_N}{E_E} = \tan^{-1}\frac{12.6}{17.4} = 35.9°.$$ The field is

$$E = \boxed{21.5 \text{ N/C at } 35.9° \text{ N of E}}.$$

16.39 (a) The field lines are sketched below.

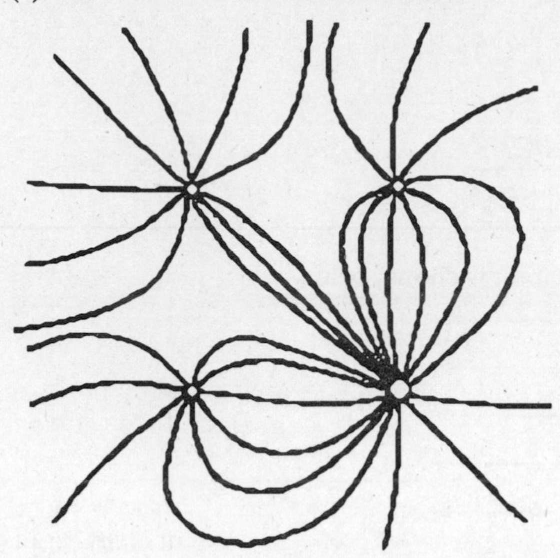

The field lines are directed away from the three positive charges and toward the larger negative charge shown in the lower right of the figure.
(b) The physical situation and the field vectors at the position of the negative charge are shown in the diagram below.

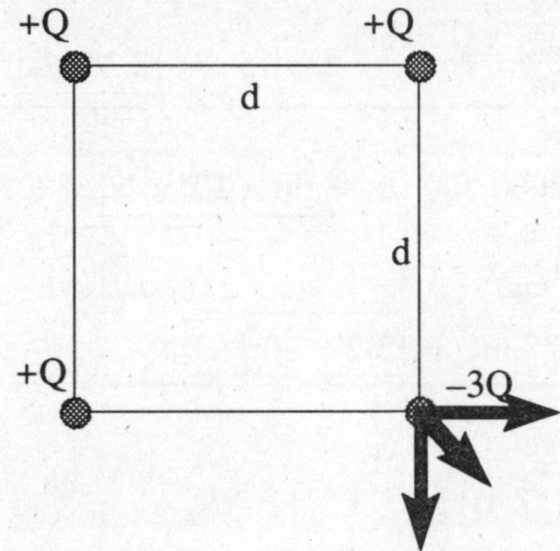

From the symmetry, the fields add to produce a net field along the diagonal. The magnitude of the field is

$$E = kQ\left(\frac{2}{d^2}\frac{1}{\sqrt{2}} + \frac{1}{(\sqrt{2}\,d)^2}\right) = kQ\left(\frac{\sqrt{2}}{d^2} + \frac{1}{(\sqrt{2}\,d)^2}\right) = \frac{kQ}{d^2}\left(\sqrt{2} + \tfrac{1}{2}\right).$$

$$\mathbf{E} = \boxed{1.91\,\frac{kQ}{d^2}\ \text{outward along the diagonal.}}$$

(c) $\mathbf{F} = q\mathbf{E} = (-3Q)\mathbf{E} = \boxed{5.74\,\frac{kQ^2}{d^2}\ \text{toward the center of the square.}}$

16.41 A sketch of the situation is shown below.

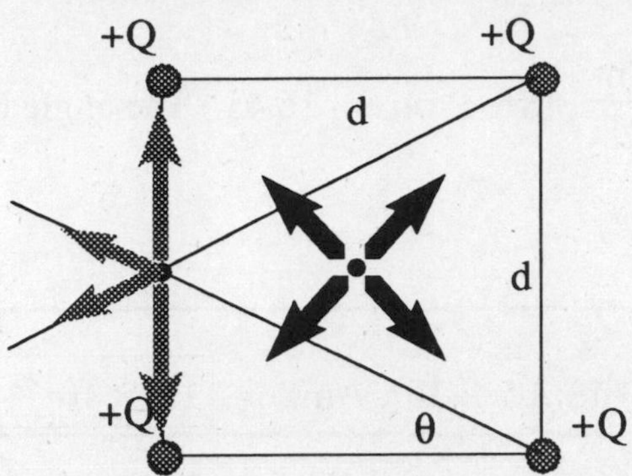

(a) At the center of the square the fields are all of the same magnitude and each points along the diagonal in such a way the the sum of the fields is zero. $\boxed{E = 0.}$

(b) At the center of one edge, the fields of the two nearest charges cancel each other so that the net field is only due to the two more distant charges. In the figure above the net field is along the horizontal direction to the left. The magnitude of the field is

$$E = 2\left[\frac{kQ}{r^2}\cos\theta\right]. \quad \text{Here } \cos\theta = \frac{d}{r}, \text{ where } r = \sqrt{d^2 + (d/2)^2} = \frac{\sqrt{5}\,d}{2}.$$

$$E = 2\left[\frac{kQ}{(5/4)d^2}\frac{2}{\sqrt{5}}\right],$$

$$E = \frac{16}{5\sqrt{5}}\frac{(9.0 \times 10^9\ \text{N·m}^2/\text{C}^2)(1.0 \times 10^{-9}\ \text{C})}{(1.0 \times 10^{-2}\ \text{m})^2} = \boxed{1.3 \times 10^5\ \text{N/C}}.$$

16.43 Inside the hollow conductor the charge is 0. From the (spherical) symmetry, any field, if present should be radial. Apply Gauss's law:

$$\phi = EA = \frac{q}{\varepsilon_o} = 0. \quad \boxed{E = 0.}$$

In the absence of high symmetry, we extend the Gaussian surface to be just inside the conductor. There the field must be zero. Since there is no charge enclosed within the Gaussian surface and the field is zero on the surface, the field inside must be zero everywhere.

16.45 Use a cylindrical Gaussian surface oriented with its axis perpendicular to the sheet and extending through it. Let area of each end be A. For a large sheet, the symmetry demands that the field be perpendicular to the sheet. Thus the flux becomes

$$\phi = \phi_{end1} + \phi_{end2} + \phi_{side} = EA + EA + 0 = \frac{q_{enclosed}}{\varepsilon_o} = \frac{\sigma A}{\varepsilon_o}.$$

$$E = \boxed{\frac{\sigma}{2\varepsilon_o}}.$$

16.47 Refer to the figure in the text. $\tan\theta = \frac{F_e}{F_g} = \frac{qE}{mg}$.

The field due to the sheet is $E = \frac{\sigma}{2\varepsilon_o}$. (See Problem 16.45.) The angle becomes

$$\theta = \boxed{\tan^{-1}\left(\frac{q\sigma}{2\varepsilon_o mg}\right)}.$$

16.49 $\Delta PE = 2\,pE = 2(3.2 \times 10^{-30}\ \text{C·m})(2.5 \times 10^3\ \text{V/m}) = \boxed{1.6 \times 10^{-26}\ \text{J}}$.

16.51 $\tau = pE = qdE$,

$$d = \frac{\tau}{qE} = \frac{1.30 \times 10^{-5}\ \text{N·m}}{3.00 \times 10^{-6}\ \text{C} \times 1.00 \times 10^3\ \text{N/C}} = \boxed{4.33\ \text{mm}}.$$

16.53 $\Delta PE = 2\,p\,E$.

$$p = \frac{\Delta PE}{2E} = \frac{2.28 \times 10^{-25}\ \text{J}}{2(1.50 \times 10^4\ \text{N/C})} = \boxed{7.60 \times 10^{-30}\ \text{C·m}.}$$

16.55 Consider the following diagram.

–q +q
a a
r

Choose r along the direction of the dipole centered at r = 0. The dipole is make of charges $\pm$ q separated by a distance 2a. The field is

$$E = kq\left(\frac{1}{(r-a)^2} - \frac{1}{(r+a)^2}\right)$$

$$E = \frac{kq}{r^2}\left[\left(1 - \frac{a}{r}\right)^{-2} - \left(1 + \frac{a}{r}\right)^{-2}\right].$$ When a<<r, we can approximate using the binomial theorem: $(1+x)^n = 1 + nx + n(n-1)x^2 + n(n-1)(n-2)x^3 \ldots$, for x<<1. If we keep only the lead terms in a/r, the result is

$$E = \frac{kq}{r^2}\left[1 + \frac{2a}{r} - 1 + \frac{2a}{r}\right] = \frac{2\,k\,2qa}{r^3} = \frac{2k\,p}{r^3} = \boxed{\frac{p}{2\pi\varepsilon_o r^3}}.$$

16.57 When the field is present, the net force is $F = qE - mg$. If the field is $E = 2mg/q$, then the net force is mg upward and a = g. Since the magnitude of the acceleration is the same, the time to go the same distance must also be the same.

16.59 (a) The two positive charges repel the test charge with equal and oppositely directed forces. The two negative charges attract the test charge with equal and oppositely directed forces. As a result the net force on the test charge is $\boxed{\text{zero}}$.

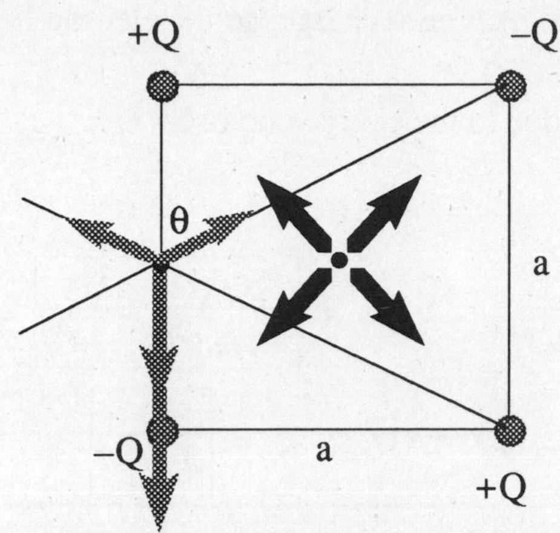

(b) Consider the test charge at the mid point of the left side as shown above. The two nearest charges each give a downward force of $F_1 = k\dfrac{Qq_o}{(a/2)^2} = \dfrac{4\,kQq_o}{a^2}$.
The magnitude of the force from each of the more distant charges is

$F_2 = k\dfrac{Qq_o}{r^2}$, where $r^2 = a^2 + (a/2)^2 = 5a^2/4$.

$F_2 = \dfrac{4\,kQq_o}{5a^2}$.

The horizontal components of the forces F_2 are equal and oppositely directed, so they cancel each other. Their vertical components add together to give an upward net force. The vertical component of each is

$$F_{2y} = F_2 \cos\theta = F_2\frac{a/2}{\sqrt{5a^2/4}} = \frac{F_2}{\sqrt{5}}.$$

Taking up as + and down as –, the net force on the test charge is

$$F_{nct} = -2\,F_1 + 2\,F_{2y} = -\frac{8\,kQq_o}{a^2} + \frac{8\,kQq_o}{5\sqrt{5a^2}} = -\frac{8\,kQq_o}{a^2}\left(1 - \frac{1}{5\sqrt{5}}\right).$$

$|F_{net}| = 7.28\,kQq_o/a^2$, so

$F_{net} = \boxed{7.28\,kQq_o/a^2 \text{ toward the nearest negative charge}}$.

16.61 (a) The magnitude of the charges is the same so that the magnitude of the force is

$$F = k\frac{q^2}{r^2} = 8.99 \times 10^9\ \text{N·m}^2/\text{C}^2\ \frac{(1.60 \times 10^{-19}\ \text{C})^2}{(5.29 \times 10^{-11}\ \text{m})^2},$$

$F = \boxed{8.22 \times 10^{-8}\ \text{N; attractive}}$.

(b) We can find the speed of the electron from the equation for centripetal force:

$$\frac{mv^2}{r} = F. \qquad v = \sqrt{\frac{Fr}{m}},$$

$$v = \sqrt{\frac{8.22 \times 10^{-8}\ \text{N} \times 5.29 \times 10^{-11}\ \text{m}}{9.11 \times 10^{-31}\ \text{kg}}} = \boxed{2.19 \times 10^6\ \text{m/s}}.$$

(c) $f = \dfrac{v}{2\pi r} = \boxed{6.57 \times 10^{15}\ \text{Hz}}$.

16.63 (a) For $r < a$, the Gaussian surface encloses zero charge, so the electric field is $\boxed{\text{zero}}$.

(b) For r between a and b the flux is due to the charge enclosed.

$$\phi = 4\pi r^2 E = \frac{\rho V}{\varepsilon_o}.$$

$$V = \frac{4}{3}\pi(r^3 - a^3) \text{ and } \rho = \frac{Q}{\frac{4}{3}\pi(b^3 - a^3)}. \qquad E = \boxed{\frac{Q(r^3 - a^3)}{4\pi\varepsilon_o r^2(b^3 - a^3)}}.$$

(c) For $r > b$, the flux is due to the total charge Q, so $E = \boxed{\frac{Q}{4\pi\varepsilon_o r^2}}$.

16.65 From the symmetry, the force on a charge at the center of a cube due to identical charges on the corners of one face of the cube must be along a line directed from the center to the mid point of the face. For a negative charge in the center and positive charges on the corners, the force will be directed outward toward the center of the face containing the positive changes. The force due to one charge alone is $F = \frac{kQq}{r^2}$. The distance r is one half the body diagonal $= \frac{\sqrt{3}}{2}a$, where a is the length of the cube edge. The component of this force perpendicular to the face of the cube is $F/\sqrt{3}$. The net force becomes

$$F_{net} = 4\left(\frac{1}{\sqrt{3}}\right)\frac{kQq}{(3/4)a^2} = \frac{16}{3\sqrt{3}}\frac{kQq}{a^2},$$

$$F_{net} = \frac{16}{3\sqrt{3}}\frac{(9.0 \times 10^9 \text{ N}\cdot\text{m}^2/\text{C}^2)(3.0 \times 10^{-7} \text{ C})^2}{(0.10 \text{ m})^2},$$

$$F_{net} = \boxed{0.25 \text{ N toward center of face}}.$$

Chapter 17

ELECTRIC POTENTIAL AND CAPACITANCE

17.1 The potential is found from Eq.(17.4) as

$$V = \frac{kq}{r} = \frac{(9.0 \times 10^9\ \text{N}\cdot\text{m}^2/\text{C}^2)(2.0 \times 10^{-3}\ \text{C})}{0.70\ \text{m}} = \boxed{26\ \text{MV}}.$$

17.3 The magnitude of the potential difference is obtained from Eq.(17.7):

$$V = Ed = (3 \times 10^4\ \text{V/m})(1.5 \times 10^{-3}\ \text{m}) = \boxed{45\ \text{V}}.$$

17.5 From Eq.(17.6) we see that

$$q = \frac{W}{V} = \frac{2 \times 10^{-6}\ \text{J}}{10^{-3}\ \text{V}} = \boxed{2 \times 10^{-3}\ \text{C}}.$$

17.7 The charge on the dome is spherically distributed so we can find the field and the potential as if from a point charge. The electric field a distance r from the dome is $E = k\frac{Q}{r^2}$. The electric potential at distance r is $V = k\frac{q}{r}$. From inspection of these two equations we see that $V = rE$.

$$V = (0.25\ \text{m})(3.0 \times 10^6\ \text{V/m}) = \boxed{7.5 \times 10^5\ \text{V}}.$$

17.9 $$V_B - V_A = k\left[\left(\frac{q_1}{r_{1B}} + \frac{q_2}{r_{2B}}\right) - \left(\frac{q_1}{r_{1A}} + \frac{q_2}{r_{2A}}\right)\right],$$ where $q_1 = q_2 = 10^{-6}$ C.

$$V_B - V_A = (9 \times 10^9\ \text{N}\cdot\text{m}^2/\text{C}^2)(10^{-6}\ \text{C})\left[\frac{1}{2\ \text{m}} + \frac{1}{2\ \text{m}} - \frac{1}{1\ \text{m}} - \frac{1}{3\ \text{m}}\right]$$

$$V_B - V_A = \boxed{-3\ \text{kV}}.$$

17.11 (a) The largest field will be just at the radius of the conductor. The maximum field is the breakdown strength. The field of a spherical charge is $E = k\frac{q}{r^2}$. Set E equal to the breakdown strength and solve for q.

$$q = \frac{r^2E}{k} = \frac{(0.10\ \text{m})^2(3.0 \times 10^6\ \text{V/m})}{9.0 \times 10^9\ \text{V}\cdot\text{m/C}} = \boxed{3.3\ \mu\text{C}}.$$

(b) $$V = \frac{kq}{r} = Er = (3.0 \times 10^6\ \text{V/m})(0.10\ \text{m}) = 3.0 \times 10^5\ \text{V} = \boxed{300\ \text{kV}}.$$

17.13 (a) The field of a spherical charge is $E = k\frac{q}{r^2}$. Right at the dome r = R = 0.15 m.

$$E = \frac{kq}{r^2} = \frac{(9.0 \times 10^9\ \text{V}\cdot\text{m/C})(3.0 \times 10^{-6}\ \text{C})}{(0.15\ \text{m})^2} = \boxed{1.2\ \text{MV/m}}.$$

(b) At a distance of 50 cm from the dome, r = 15 cm + 50 cm = 65 cm = 0.65 m.

$$E = \frac{kq}{r^2} = \frac{(9.0 \times 10^9\ \text{V}\cdot\text{m/C})(3.0 \times 10^{-6}\ \text{C})}{(0.65\ \text{m})^2} = \boxed{64\ \text{kV/m}}.$$

(c) The potential on the dome is $V = k\frac{q}{r}$, where r = R = 0.15 m .

$$V = \frac{(9.0 \times 10^9\ \text{V}\cdot\text{m/C})(3.0 \times 10^{-6}\ \text{C})}{0.15\ \text{m}} = \boxed{1.8 \times 10^5\ \text{V}}.$$

17.15 (a) First make a sketch of the situationl.

+q -q
0 x
0.10 m

$V = \frac{kq_1}{r_1} + \frac{kq_2}{r_2}$. Here, $q_1 = q = 1.0 \times 10^{-9}$ C and $q_2 = -1.0 \times 10^{-9}$ C.

$r_1 = |x|$ and $r_2 = |x - 0.10\ \text{m}|$. If x is in meters, then

$$V = kq\left(\frac{1}{|x|} - \frac{1}{|x - 0.10|}\right) = \boxed{9.0\ \text{V}\left(\frac{1}{|x|} - \frac{1}{|x - 0.10|}\right)}.$$

(b) Let x = – 0.050 m.

$$V = 9.0\ \text{V}\left(\frac{1}{|-0.050|} - \frac{1}{|-0.050 - 0.10|}\right)$$

$$V = 9.0\ \text{V}\,(20 - 6.67) = \boxed{120\ \text{V}}.$$

17.17 (a) The speed is obtained from the definition of kinetic energy: $KE = \frac{1}{2}mv^2$,

$$v = \sqrt{\frac{2\ KE}{m}} = \sqrt{\frac{(2)(200\ \text{eV})(1.60 \times 10^{-19}\ \text{J/eV})}{1.67 \times 10^{-27}\ \text{kg}}} = \boxed{1.96 \times 10^5\ \text{m/s}}.$$

$$\text{(b)}\ v = \sqrt{\frac{2\ KE}{m}} = \sqrt{\frac{(2)(200\ \text{eV})(1.60 \times 10^{-19}\ \text{J/eV})}{9.11 \times 10^{-31}\ \text{kg}}} = \boxed{8.38 \times 10^6\ \text{m/s}}.$$

17.19 (a) The kinetic energy is given by

$KE = qV = eV = (e)(250{,}000\ \text{V}) = 250000\ \text{eV} = \boxed{2.50 \times 10^5\ \text{eV}}$.

(b) In joules the energy is

$2.5 \times 10^5\ \text{eV} \times 1.60 \times 10^{-19}\ \text{J/eV} = \boxed{4.00 \times 10^{-14}\ \text{J}}$.

(c) The speed is obtained from the definition of kinetic energy: $KE = \frac{1}{2}mv^2$,

$$v = \sqrt{\frac{2\ KE}{m}} = \sqrt{\frac{(2)(4.00 \times 10^{-14}\ \text{J.})}{1.67 \times 10^{-27}\ \text{kg}}} = \boxed{6.92 \times 10^6\ \text{m/s}}.$$

17.21 Because the potentials are the same and the radii are the same, the charges will also be the same. Each sphere will have a charge $\boxed{Q/2}$.

17.23 (a) Since the potentials are equal, $V = \frac{kq_1}{r_1} = \frac{kq_2}{r_2}$. $q_1 = \frac{r_1}{r_2} q_2$.

The $\boxed{\text{first}}$ sphere has the greater charge since $r_1 > r_2$.

(b) $E = \frac{V}{r}$. Since the potentials are the same, the $\boxed{\text{second}}$ sphere will have the greatest field at its surface.

17.25 Initially the first sphere has a potential $V_o = 180\ V = \frac{kQ}{r_1}$.

After touching, both spheres have the same potential and the charge Q is distributed between them.

$V = \frac{kq_1}{r_1} = \frac{kq_2}{r_2} = \frac{k(Q - q_1)}{2r_1}$. Solving this equation we get

$q_1 + \frac{1}{2}q_1 = \frac{1}{2}Q$. The final potential can be computed now that we know the charge. $V = \frac{kq_1}{r_1} = \frac{1}{3}\frac{kQ}{r_1} = \frac{180\ V}{3} = \boxed{60\ V}$.

17.27 $C = \frac{q}{V} = \frac{18 \times 10^{-5}\ C}{9.0\ V} = \boxed{20\ \mu F}$.

17.29 Rearrange the definition of capacitance to get

$V = \frac{q}{C} = \frac{3.0 \times 10^{-3}\ C}{20 \times 10^{-6}\ F} = \boxed{150\ V}$.

17.31 We know that the charge is proportional to the voltage. If the voltage is doubled, then the charge is doubled (the capacitance is constant). Thus when the final voltage is 30 V, the final charge is $2 \times 20\ \mu C = 40\ \mu C$. The capacitance is

$C = \frac{q}{V} = \frac{40 \times 10^{-6}\ C}{30\ V} = \boxed{1.33\ \mu F}$.

17.33 $C = \frac{\varepsilon_o A}{d} = \frac{(8.85 \times 10^{-12}\ F/m)(1.13 \times 10^{-2}\ m^2)}{1.0 \times 10^{-4}\ m} = \boxed{1.0\ nF}$.

17.35 $q = CV = \frac{\varepsilon_o AV}{d} = \frac{(8.85 \times 10^{-12}\ F/m)\pi(0.25\ m)^2(750\ V)}{2.5 \times 10^{-3}\ m} = \boxed{0.52\ \mu C}$.

17.37 $W = Fx = qEx = (1.6 \times 10^{-19}\ C)(7.0\ V/m)(0.010\ m) = \boxed{1.1 \times 10^{-20}\ J}$.

17.39 Use the definition of capacitance and Eq.(17.10) for the capacitance of a parallel plate capacitor.

$$q = CV = \frac{\varepsilon_o A}{d} V = \frac{(8.85 \times 10^{-12}\ \text{F/m})\pi\left(\frac{0.185\ \text{m}}{2}\right)^2}{2.00 \times 10^{-3}\ \text{m}} (15.7\ \text{V}),$$

$q = \boxed{1.87 \times 10^{-9}\ \text{C}}$.

17.41 Imagine a small cylindrical Gaussian surface placed between the plates and oriented with its axis perpendicular to the plates.

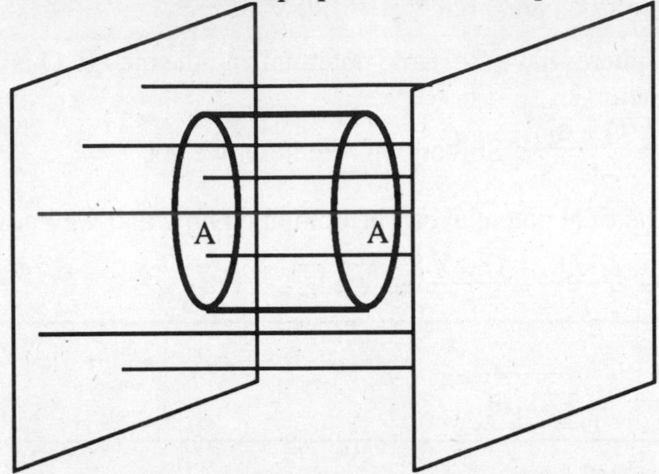

The electric field is parallel to the axis of the cylinder and to the cylindrical surface, so there is no flux through that surface. There is no net charge contained within the surface of the cylinder, so the net flux through it must be zero. Consequently, the flux entering at one end must be exactly equal to the flux leaving at the other end. Since the surface areas of the two ends are identical, the electric field at either end must also be identical. Because we did not require the Gaussian surface to be any particular size, we can conclude that the electric field E must be the same (constant) everywhere between the plates.

17.43 The maximum field that the insulator can withstand is the breakdown strength. Using the relationship between field, potential, and distance we get

$d = \frac{V}{E} = \frac{25\ \text{V}}{526 \times 10^6\ \text{V/m}} = \boxed{4.8 \times 10^{-8}\ \text{m}}$.

17.45 For a parallel plate capacitor the capacitance is given by $C = \frac{\kappa\varepsilon_o A}{d}$. Upon

rearranging we get $A = \frac{Cd}{\kappa\varepsilon_o} = \frac{(10 \times 10^{-6}\ \text{F})(2.0 \times 10^{-8}\ \text{m})}{8.0(8.85 \times 10^{-12}\ \text{F/m})}$,

$A = \boxed{2.8 \times 10^{-3}\ \text{m}^2}$.

17.47 $C = \frac{\kappa\varepsilon_o A}{d}$ and $V = Ed$. Their product is

$CV = \boxed{\kappa\varepsilon_o AE}$, where E is the dielectric strength.

17.49 (a) $C = \frac{\kappa\varepsilon_o A}{d} = \frac{(11)(8.85 \times 10^{-12}\text{ F/m})(0.60\text{ m}^2)}{25 \times 10^{-6}\text{ m}} = \boxed{2.3\ \mu\text{F}}$.

(b) $V = Ed = (200 \times 10^6\text{ V/m})(25 \times 10^{-6}\text{ m}) = \boxed{5\text{ kV}}$.

17.51 (a) $q = CV = 0.010 \times 10^{-6}\text{ F} \times 10\text{ V} = \boxed{0.10\ \mu\text{C}}$.

(b) $W = \frac{1}{2}CV^2 = \frac{1}{2}(0.010 \times 10^{-6}\text{ F})(10\text{ V})^2 = \boxed{5.0 \times 10^{-7}\text{ J}}$.

17.53 (a) First find C_1, the capacity with the glass plate in place.

$$C_1 = \frac{\kappa\varepsilon_o A}{d} \frac{(6.3)(8.85 \times 10^{-12}\text{ F/m})(0.30\text{ m}^2)}{3.0 \times 10^{-3}\text{ m}} = 1.67 \times 10^{-9}\text{ F}.$$

$$W_1 = \frac{1}{2}C_1V_1^2 = \frac{1}{2}(1.67 \times 10^{-9}\text{ F})(500\text{ V})^2 = \boxed{2.1 \times 10^{-4}\text{ J}}.$$

(b) Without the glass the capacitance becomes $C_2 = C_1/\kappa$. The charge remains because the capacitor was insulated from surroundings so $V_2 = q/C_2 = \kappa V_1$.

$$W_2 = \frac{1}{2}C_2V_2^2 = \frac{1}{2}\frac{C_1}{\kappa}(\kappa V_1)^2 = \kappa W_1 = \boxed{1.3 \times 10^{-3}\text{ J}}.$$

Notice that the work done to remove the glass results in an increase in the electric potential energy of the capacitor.

17.55 Let us choose the positive direction of d to be in the direction of increasing potential, that is, from the lower-voltage plate to the higher- voltage one. We can then apply Eq. (17.7): $E = -\frac{V}{d}$.

Here $V = +300$ V when $d = +0.75$ cm $= 0.75 \times 10^{-2}$ m. Thus

$$E = -\frac{300\text{ V}}{0.75 \times 10^{-2}\text{ m}} = -40{,}000\text{ V/m}.$$

$\boxed{\text{E has a magnitude of 40 000 V/m directed to plate with lower potential.}}$

17.57 (a) The kinetic energy is $KE = \frac{1}{2}mv^2 = |qV|$. The speed is then given by

$$v = \sqrt{\frac{2|qV|}{m}} = \sqrt{\frac{2(1.6 \times 10^{-19}\text{ C})(750\text{ V})}{9.1 \times 10^{-31}\text{ kg}}} = \boxed{1.6 \times 10^7\text{ m/s}}.$$

(b) The KE is $|qV| = eV$, where the charge is $-e$ and the potential is 750 volts.

$KE = \boxed{750\text{ eV}}$.

17.59 Let d represent the length of a side.

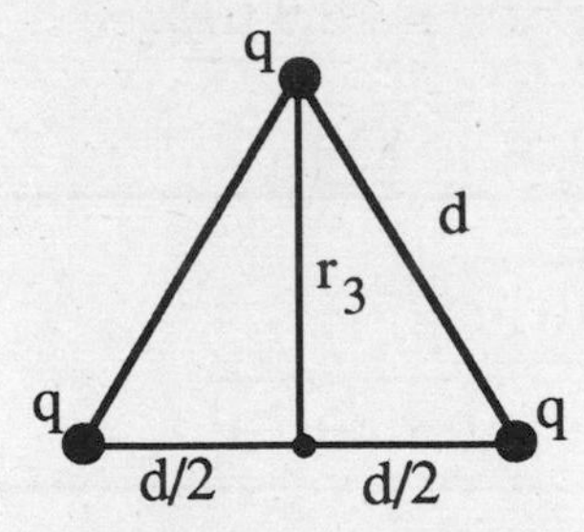

The distance to the nearest charges is $r_1 = r_2 = d/2$. The distance to the third charge is obtained from the Pythagorean theorem:

$$r_3^2 + \left(\frac{d}{2}\right)^2 = d^2. \quad r_3 = \sqrt{\frac{3}{4}d^2} = 0.87\ d.$$

$$V = kq\left(\frac{1}{r_1} + \frac{1}{r_2} + \frac{1}{r_3}\right) = \frac{kq}{d}\left(2 + 2 + \frac{1}{0.87}\right) = 5.15\frac{kq}{d},$$

$$V = 5.15\frac{(9.0 \times 10^9\ \text{V·m/C})(1.0 \times 10^{-6}\ \text{C})}{1.0 \times 10^{-2}\ \text{m}} = \boxed{4.6 \times 10^6\ \text{V}}.$$

17.61 (a) In cross section we show a Gaussian surface passing through a charged sheet.

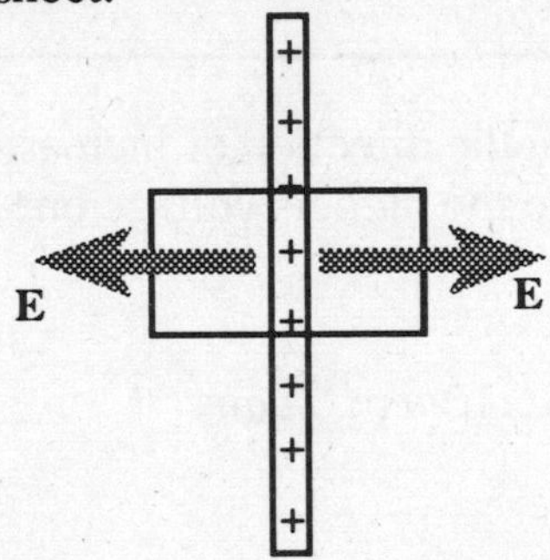

From Gauss's law, $EA_{left} + EA_{right} = \frac{q}{\varepsilon_o} = \frac{\sigma A}{\varepsilon_o}$.

$$2EA = \frac{\sigma A}{\varepsilon_o}.$$

$$E = \frac{\sigma}{2\varepsilon_o} = \frac{3.54 \times 10^{-8}\ \text{C/m}^2}{2 \times 8.85 \times 10^{-12}\ \text{F/m}} = \boxed{2.00\ \text{kV/m}}.$$

(b) Use the relation $|E| = \frac{V}{d}$.

$$2000\ \text{V/m} = \frac{100\ \text{V}}{d}.$$

$$d = \frac{100\ \text{V}}{2000\ \text{V/m}} = 0.050\ \text{m} = \boxed{5.0\ \text{cm}}.$$

17.63 With polyethylene in place $C_o = \dfrac{\kappa\varepsilon_0 A}{d}$. With the plastic removed the capacitance becomes $C = \dfrac{\varepsilon_0 A}{d}$. So $C = C_o/\kappa$. Since the charge q is constant, $V = \kappa V_o$.
Work done is the difference between the final and initial stored energies

$$\Delta W = \frac{1}{2}CV^2 - \frac{1}{2}C_oV_o^2 = \frac{1}{2}\frac{C_o}{\kappa}(\kappa V_o)^2 - \frac{1}{2}C_oV_o^2 = \boxed{\frac{1}{2}C_oV_o^2(\kappa - 1)}.$$

17.65 (a) $W_o = \frac{1}{2}C_oV_o^2 = \frac{1}{2}(100 \times 10^{-6}\text{ F})(500\text{ V})^2 = \boxed{12.5\text{ J}}$.

(b) $V \propto d$, so that $V = 2V_o$. Also $C = C_o/2$.

$$W = \frac{1}{2}CV^2 = \frac{1}{2}\frac{C_o}{2}(2v_o)^2 = 2\,W_o = \boxed{25\text{ J}}.$$

(c) Work must be supplied to separate the plates.

Chapter 18

ELECTRIC CURRENT AND RESISTANCE

18.1 From the definition of electric current we get

$$I = \frac{\Delta q}{\Delta t} = \frac{1800\ \text{C}}{(1.00\ \text{min})(60.0\ \text{s/min})} = \boxed{30.0\ \text{A}}.$$

18.3 Use the definition of electric current to find

(a) $\Delta q = I\,\Delta t = (0.50\ \text{A})((1.0\ \text{s}) = \boxed{0.50\ \text{C}}$.

(b) $\Delta q = (0.50\ \text{A})(60\ \text{s}) = \boxed{30\ \text{C}}$.

18.5 $$I = \frac{\Delta q}{\Delta t} = \frac{(250 \times 10^{6})(1.6 \times 10^{-19}\ \text{C})}{0.010\ \text{s}} = 4.0 \times 10^{-9}\ \text{A} = \boxed{4\ \text{nA}}.$$

18.7 From Ohm's law we get

$V = IR = (3.00\ \text{A})(330\ \Omega) = \boxed{990\ \text{V}}$.

18.9 Using the definition of resistance we get

$$R = \frac{V}{I} = \frac{3(1.0\ \text{V})}{0.60\ \text{A}} = \boxed{5.0\ \Omega}.$$

18.11 Using the definition of resistance we get

$$R = \frac{V}{I} = \frac{12\ \text{V}}{(8.0 \times 10^{4}\ \text{C/h})/(3600\ \text{s/h})} = \boxed{0.54\ \Omega}.$$

18.13 From the graph, $R = 200\ \Omega$ for all values of V.
The current is obtained from $I = V/R$, which gives, beginning at 0 V, in increment of 20 V:

$I = \boxed{0.00\ \text{A; } 0.10\ \text{A; } 0.20\ \text{A; } 0.30\ \text{A; } 0.40\ \text{A; } 0.50\ \text{A; } 0.60\ \text{A}}$.

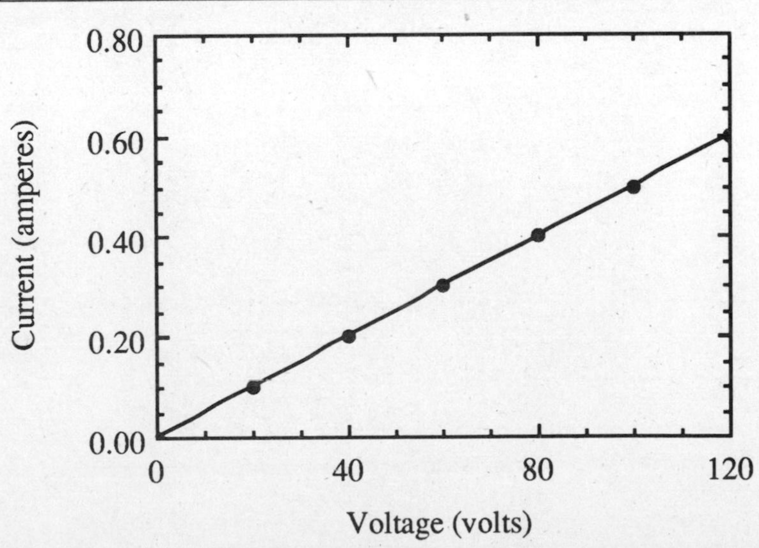

18.15 From Eq. (18.4) and Table 18.1 we have

$$R = \frac{\rho L}{A} = \frac{(1.72 \times 10^{-8}\ \Omega\cdot m)(2.00\ m)}{0.5176 \times 10^{-6}\ m^2} = \boxed{6.65 \times 10^{-2}\ \Omega}.$$

18.17 From Eq. (18.4) and Table 18.1 we have

$$R = \frac{\rho L}{A} = \frac{(9.7 \times 10^{-8}\ \Omega\cdot m)(2.00\ m)}{0.90 \times 10^{-6}\ m^2} = \boxed{0.22\ \Omega}.$$

18.19 From Eq. (18.4) we have $R = \frac{\rho L}{A}$.

$$R_{(18)} = \frac{\rho L}{A} = \frac{(1.72 \times 10^{-8}\ \Omega\cdot m)(5.00\ m)}{0.8231 \times 10^{-6}\ m^2} = \boxed{0.104\ \Omega}.$$

$$R_{(16)} = \frac{\rho L}{A} = \frac{(1.72 \times 10^{-8}\ \Omega\cdot m)(5.00\ m)}{1.309 \times 10^{-6}\ m^2} = \boxed{0.0657\ \Omega}.$$

18.21 From Eq.(18.5) we have $P = IV = (0.10\ A)(1.4\ V) = \boxed{0.14\ W}$.

18.23 From Eq.(18.5) we have $I = \frac{P}{V} = \frac{800\ W}{120\ V} = \boxed{6.67\ A}$.

18.25 Cost = Ptr = (0.840 kW)(8 h)($0.080/kWh) = $\boxed{\$\,0.54}$.

18.27 Cost = Ptr = (0.100 kW)(8 h/d)(30 d)($ 0.075/kWh) = $\boxed{\$\,1.80}$.

18.29 (a) $P = \frac{V^2}{R}$. $V = \sqrt{PR} = \sqrt{(1.00\ W)(100\ \Omega)} = \boxed{10.0\ V}$.

(b) $I = \frac{V}{R} = \frac{10.0\ V}{100\ \Omega} = \boxed{0.100\ A}$.

18.31 (a) When the switch is open the same current passes through each resisitor. Their combined reistance is 200 Ω. $I = \frac{V}{R} = \frac{9.0\ V}{200\ \Omega} = \boxed{0.045\ A}$.

(b) When the switch is closed, the current bypasses the left resistor so the current in the left resistor is $\boxed{\text{zero}}$ and the current in the right resistor is given by

$$I = \frac{V}{R} = \frac{9.0\ V}{100\ \Omega} = \boxed{0.090\ A}.$$

18.33 $$R_T = \frac{R_1 R_2}{R_1 + R_2} = \frac{100\ \Omega \times 33\ \Omega}{133\ \Omega} = \boxed{25\ \Omega}.$$

18.35 First find the effective resistance of the three 390-Ω resistors in parallel.

$$\frac{1}{R_P} = \frac{1}{390\ \Omega} + \frac{1}{390\ \Omega} + \frac{1}{390\ \Omega} = \frac{3}{390\ \Omega}. \qquad R_P = 130\ \Omega.$$

Now add the two remaining resistors in series with R_P.

$$R_T = R_P + 150\ \Omega + 120\ \Omega = \boxed{400\ \Omega}.$$

18.37 (a) First find the total resistance R_T. $R_T = 3(47\ \Omega) + 15\ \Omega = 156\ \Omega$.
The current through each of the resistors is the same. It is

$$I = \frac{V_{battery}}{R_T} = \frac{9.0\ V}{156\ \Omega} = 5.77 \times 10^{-2}\ A = \boxed{58\ mA}.$$

(b) $V = IR = (58 \times 10^{-3}\ A)(15\ \Omega) = \boxed{0.87\ V}$.

18.39 Use the results of Problem 18.38.

$$R_x = \frac{R_2R_3}{R_1} = \frac{(20.00\ \Omega)(18.31\ \Omega)}{100.0\ \Omega} = \boxed{3.662\ \Omega}.$$

18.41 Capacitors in series add according to Eq. (18.12).

$$\frac{1}{C_T} = \left(\frac{1}{1.0} + \frac{1}{1.5} + \frac{1}{5.0}\right)\frac{1}{\mu F} = \frac{28}{15\ \mu F}, \qquad C_T = \boxed{0.54\ \mu F}.$$

18.43 Capacitors in parallel add directly according to Eq. (18.11).

$$C_T = C_1 + C_2 + C_3 = \boxed{7.5\ \mu F}.$$

18.45 For each pair of plates:

$$C = \frac{\varepsilon_0 A}{d} = \frac{(8.85 \times 10^{-12}\ F/m)(6.0 \times 10^{-4}\ m^2)}{5.0 \times 10^{-4}\ m} = 1.06 \times 10^{-11}\ F.$$

Because these are all in parallel, their combination is a capacitance

$$C_T = 15\ C = 1.6 \times 10^{-10}\ F = \boxed{160\ pF}.$$

18.47 Consider the loop containing the battery and the two series capacitors. The total voltage is applied across the series capacitors. They each must store the same charge. That charge is $q = CV$, where C is their series capacitance,

$$C = \frac{(10)(20)}{10 + 20}\ \mu F = \frac{20}{3}\ \mu F.$$

So, $q = \frac{20}{3}\ \mu F\ (9.0\ V) = 60\ \mu C$. The voltage across the 20 μF capacitor is

$$V = \frac{q}{C} = \frac{60\ \mu C}{20\ \mu F} = \boxed{3.0\ V}.$$

18.49 There are $\boxed{\text{seven combinations}}$. They are: one alone, two in series, two in parallel, three in series, three in parallel, one in series with two in parallel, and one in parallel with two in series. The values arranged in order of increasing capacitance are $\boxed{5\ \mu F;\ 7.5\ \mu F;\ 10\ \mu F;\ 15\ \mu F;\ 22.5\ \mu F;\ 30\ \mu F;\ 45\ \mu F}$.

18.51 (a) $V_1 + V_2 = 12\ V$. $q_1 = C_1V_1 = C_2V_2 = q_2$.

$C_1 = 15\ \mu F,\ C_2 = 25\ \mu F$.

$$V_1 = \frac{C_2}{C_1}V_2 = \frac{C_2}{C_1}(V - V_1),$$

$$V_1 = \frac{25}{15}(12\ V - V_1),$$

$$\frac{8}{3}V_1 = \frac{5}{3}(12\ V),$$

$$V_1 = \frac{5}{8}(12\ V) = \boxed{7.5\ V}.$$

$$V_2 = 12\ V - V_1 = 12\ V - 7.5\ V = \boxed{4.5\ V}.$$

(b) $q_2 = q_1 = C_1V_1 = 15\ \mu F \times 7.5\ V = \boxed{112.5\ \mu C}$.

(c) $W_1 = \frac{1}{2}C_1V_1^2 = \dfrac{(15\ \mu F)(7.5\ V)^2}{2} = \boxed{4.2 \times 10^{-4}\ J}$.

$$W_2 = \frac{1}{2}C_2V_2^2 = \frac{(25\ \mu F)(4.5\ V)^2}{2} = \boxed{2.5 \times 10^{-4}\ J}.$$

18.53 The batteries are joined in series so the total emf is $3 \times 1.5\ V = 4.5\ V$.
The definition of internal resistance r gives

$TPD = \mathscr{E} - Ir$, so

$$r = \frac{\mathscr{E} - TPD}{I} = \frac{4.5\ V - 2.8\ V}{0.60\ A} = 2.83\ \Omega.$$

We have just computed the total internal resistance. Assuming each cell has the same resistance we find that each cell has a resistance of $\boxed{0.94\ \Omega}$.

18.55 (a) Using Eq. (18.13) we find that

$$TPD = \frac{R_L\mathscr{E}}{R_L + r} = \frac{(1500\ \Omega)(9.0\ V)}{1875\ \Omega} = \boxed{7.2\ V}.$$

(b) For the new value of the load resistance we get

$$TPD = \frac{(750\ \Omega)(9.0\ V)}{1125\ \Omega} = \boxed{6.0\ V}.$$

(c) $P = \dfrac{V^2}{R} = \dfrac{(TPD)^2}{R} = \dfrac{(6.0\ V)^2}{750\ \Omega} = 0.048\ W = \boxed{48\ mW}$.

18.57 $P = IV = (15\ A)(120\ V) = 1800\ W$.
The circuit can carry $\boxed{18}$ one-hundred watt lamps.

18.59 $P_{max} = I_{max}V = 20\ A \times 120V = 2400\ W.$

(a) $\boxed{\text{No}}$, they cannot be operated simultaneously.

(b) Only the $\boxed{\text{lamp can be operated with either the toaster or the iron}}$.

18.61 $R_T = R_{Cu} + R_{Fe} = \dfrac{\rho_{Cu}L_{Cu} + \rho_{Fe}L_{Fe}}{A},$

$$R_T = \frac{[1.72(0.370\ m) + 9.71(0.185\ m)]\ 10^{-8}\ \Omega\cdot m}{(\pi/4)(2.30 \times 10^{-2}\ m)^2} = \boxed{5.86 \times 10^{-5}\ \Omega}.$$

18.63 $P = 75\ W + \left(\dfrac{1}{256\ hp}\right)(746\ W/hp) = 75\ W + 2.9\ W = 77.9\ W.$

Cost = power × time × rate, so the time is

$$t = \frac{\text{cost}}{\text{power} \times \text{rate}} = \frac{1\ ¢}{0.0779\ kW \times 7.6¢/kWh} = \boxed{1.7\ h}.$$

18.65 (a) $\boxed{\text{Zero}}$, because the 10-Ω resistor is shorted by the wire across its ends.

(b) First find the resistance equivalent to the parallel combination of the 20-Ω and the 5.0-Ω resistors.

$$R_p = \frac{20 \times 5.0}{20 + 5.0}\ \Omega = 4.0\ \Omega.$$

Next find the voltage across R_p.

$$V_p = \frac{R_p}{R_p + 5.0\ \Omega} V = \frac{4.0\ \Omega}{4.0\ \Omega + 5.0\ \Omega} 12V = 5.3\ V.$$

Thus the 20-Ω resistor has a potential difference of 5.33 V across it, which means that the current through the resistor is

$$I = \frac{5.3\ V}{20\ \Omega} = \boxed{0.27\ A}.$$

18.67 $V_A = \dfrac{R_2}{R_1 + R_2} V$, and $V_C = \dfrac{R_x}{R_x + R_3} V.$

$$V_A - V_C = \left(\frac{5.00}{10.0 + 5.00} - \frac{9.00}{9.00 + 20}\right) 12.0\ V = \boxed{0.276\ V}.$$

18.69 First notice that in the left-hand loop the two batteries are oppositely directed, so the net emf in that loop is 9.0 V - 1.5 V = 7.5 V.
The other resistors can be added together to find the potential from A to B.
The equivalent circuit is:

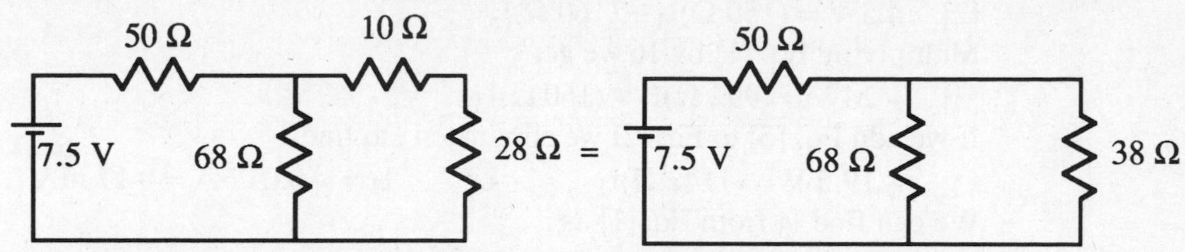

The 68-Ω resistor can be combined with the 38-Ω resistance to give 24.4 Ω.

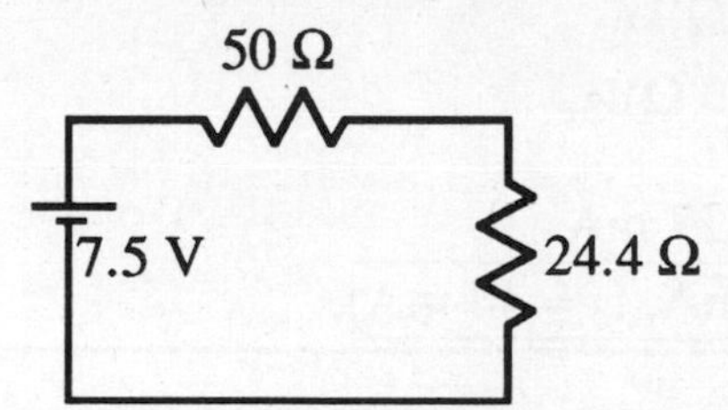

The voltage across the 24.4-Ω resistance is

$$V = \frac{24.4\ \Omega}{24.4\ \Omega + 50\ \Omega} 7.5\ \text{V} = 2.46\ \text{V}.$$

This voltage is the potential difference from A to B and is the potential across the 38-Ω resistance. Thus the current in that branch is

$$I = \frac{2.46\ \text{V}}{38\ \Omega} = 0.0647\ \text{A} = 64.7\ \text{mA}.$$

From the circuit diagram, we see that all of this current passes through the 10-Ω resistor. It is appropriate here to round off the answer to two significant figures giving the current through the 10-Ω resistor as $\boxed{65\ \text{mA}}$.

18.71 Starting with Kirchhoff's current rule we see from the diagram that $I_1 + I_3 = I_2$.

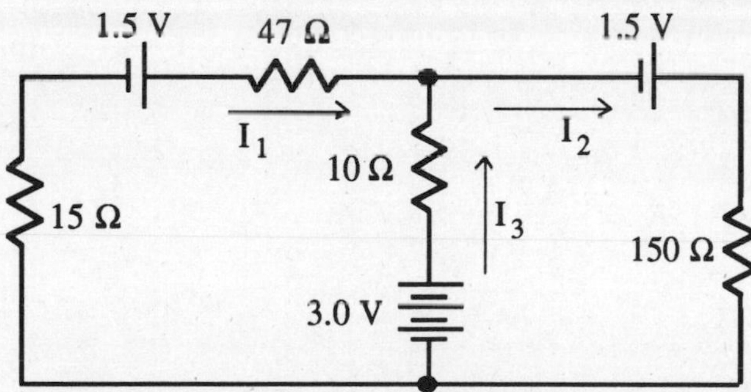

Writing the loop rule for a clockwise traversal of the left-hand loop gives

$$1.5\ \text{V} - 3.0\ \text{V} - (62\ \Omega)I_1 + (10\ \Omega)I_3,$$

[1] $-1.5\ \text{V} = (6.2\ \Omega)I_1 - (10\ \Omega)I_3.$

Writing the loop rule for clockwise traversal of the right-hand loop gives

$$1.5\ \text{V} + 3.0\ \text{V} - (150\ \Omega)I_2 - (10\ \Omega)I_3,$$

$$4.5\ \text{V} = +\ (150\ \Omega)I_2 + (10\ \Omega)I_3,$$

Using the current equation to eliminate I_2, we get

$$4.5\ \text{V} = (150\ \Omega)(I_1 + I_3) + (10\ \Omega)I_3$$

[2] $4.5 \text{ V} = (150 \ \Omega)I_1 + (160 \ \Omega)I_3$.

Multiplying Eq. [1] by 16 we get

[3] $-24 \text{ V} = (992 \ \Omega)I_1 - (160 \ \Omega)I_3$.

If we add Eq. [3] to Eq. [2] we eliminate I_3 to find

$$-19.5 \text{ V} = (1142 \ \Omega)I_1 \quad \text{or} \quad I_1 = -0.017 \text{ A} = -17 \text{ mA}.$$

We can find I_3 from Eq. [1] as

$$-1.5 \text{ V} = (62 \ \Omega)(-0.017 \text{ A}) - (10 \ \Omega)I_3,$$

$$-1.5 \text{ V} + (62 \ \Omega)(0.017 \text{ A}) = -(10 \ \Omega)I_3,$$

$$-1.5 \text{ V} + 1.06 \text{ V} = -0.44 \text{ V} = -(10 \ \Omega)I_3,$$

$$I_3 = 0.044 \text{ A} = 44 \text{ mA}$$

Then $I_2 = I_1 + I_3 = -17 \text{ mA} + 44 \text{ mA} = 27 \text{ mA}$.

The currents are $\boxed{I_1 = -17 \text{ mA}, I_2 = 27 \text{ mA}, I_3 = 44 \text{ mA}}$.

Chapter 19

MAGNETISM

19.1

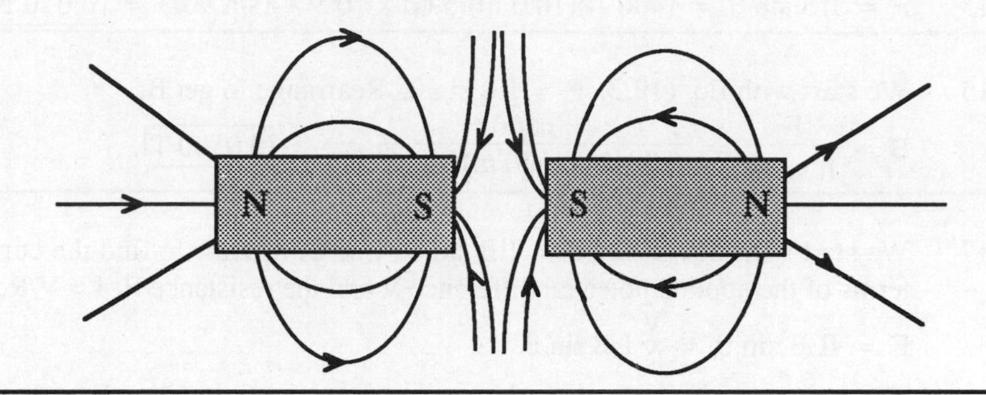

19.3 Eq. (19.1) relates the torque to the dipole moment and the magnetic field:

$\tau = \mu B \sin\theta$, which can be rearranged to give

$\mu = \dfrac{\tau}{B \sin\theta}$. The corresponding equation for the units is

$$[\mu] = \frac{[\tau]}{[B][\sin\theta]} = \frac{[\tau]}{[B]} = \frac{N\cdot m}{N/A\cdot m} = \boxed{A\cdot m^2}.$$

19.5 We rearrange Eq. (19.1) to get

$$\mu = \frac{\tau}{B \sin\theta} = \frac{0.050\ N\cdot m}{(0.30T)(\sin 45°)} = \boxed{0.24\ A\cdot m^2}.$$

19.7 The maximum torque occurs when the dipole is at right angles to the magnetic field so that θ in Eq. (19.1) is 90°.Then

$$\tau_{max} = \mu B = (7.5 \times 10^6\ A\cdot m^2)(5.00 \times 10^{-5}\ T) = \boxed{375\ N\cdot m}.$$

19.9 We can make analogy with the analysis of the electric dipole moment in an electric field that we saw in Chapter 16. There we saw that the torque on the dipole was

$\tau = pE \sin\theta$, where p was the electric dipole moment. We also found that the energy of the dipole in the field could be express as $PE = -pE\cos\theta$. By analogy we expect that the potential energy of a magnetic dipole in a magnetic field is $PE = -\mu B \cos\theta$.

Then the work done is simple the difference in the final and initial potential energies:

$$W = PE_{final} - PE_{initial} = \mu B - (-\mu B) = \boxed{2\mu B}.$$

19.11 $F = IlB. \quad I = \frac{F}{lB} = \frac{mg}{lB} = \frac{(1.0 \times 10^{-3}\text{ kg})(9.8\text{ m/s}^2)}{(1.0 \times 10^{-3}\text{ T})(0.50\text{ m})} = \boxed{20\text{ A}}.$

19.13 $F = IlB \sin\theta = (400\text{ A})(10.0\text{ m})(5.00 \times 10^{-6}\text{ T})(\sin 90°) = \boxed{0.020\text{ N}}.$

19.15 We start with Eq. (19.2), $F = IlB \sin\theta$. Rearrange to get B:

$$B = \frac{F}{Il \sin\theta} = \frac{0.106\text{ N}}{(6.00\text{ A})(0.500\text{ m})(\sin 45.0°)} = \boxed{0.0500\text{ T}}.$$

19.17 We start with Eq. (19.2), $F = IlB \sin\theta$. Use Ohm's law to find the current I in terms of the applied potential difference V and the resistance R: $I = V/R$.

$$F = ILB \sin\theta = \frac{V}{R} LB \sin\theta.$$

The resistance can be expressed in terms of the resistivity, length, and cross-sectional area.

$$F = \frac{VLB}{\rho L/A} \sin\theta = \boxed{\frac{VBA \sin\theta}{\rho}}.$$ This expression is independent of the length L.

19.19 Rearranging Eq. (19.4) we find the magnetic field to be

$$B = \frac{mv}{qr} = \frac{(6.64 \times 10^{-27}\text{ kg})(1.85 \times 10^{7}\text{ m/s})}{(3.20 \times 10^{-19}\text{ C})(0.580\text{ m})} = \boxed{0.662\text{ T}}.$$

19.21 The initial velocity v_o makes an angle of 45° with the magnetic field, thus along the direction of field there is a velocity component $v_o/\sqrt{2}$ and in a direction perpendicular to the field there is a velocity component that also has a magnitude of $v_o/\sqrt{2}$. The motion along the field direction is unaffected by the field, the velocity transverse to the field gives rise to an acceleration that causes circular motion. Thus the combined motion is $\boxed{\text{helical}}$ with a radius

$$r = \frac{mv \sin\theta}{qB} = \frac{(1.67 \times 10^{-27}\text{ kg})(1.20 \times 10^{6}\text{ m/s})(\sin 45°)}{(1.60 \times 10^{-19}\text{ C})(1.05\text{ T})}$$

$$r = \boxed{0.843\text{ cm}}.$$

19.23 (a) $F = qvB = (3.2 \times 10^{-19}\text{ C})(3.0 \times 10^{6}\text{ m/s})(0.10\text{ T}) = \boxed{9.6 \times 10^{-14}\text{ N}}.$

(b) $E = \frac{F}{q} = \frac{9.60 \times 10^{-14}\text{ N}}{3.2 \times 10^{-19}\text{ C}} = \boxed{3.0 \times 10^{5}\text{ N/C}}.$

19.25 (a) Rearrange Eq. (19.4) to get

$$v = \frac{qBr}{m} = \frac{(1.60 \times 10^{-19}\text{ C})(0.100\text{ T})(0.0500\text{ m})}{1.67 \times 10^{-27}\text{ kg}} = \boxed{4.79 \times 10^{5}\text{ m/s}}.$$

(b) $KE = \frac{1}{2}mv^2 = \frac{1}{2}(1.67 \times 10^{-27}\ kg)(4.79 \times 10^5\ m/s)^2$,

$KE = \boxed{1.92 \times 10^{-16}\ J}$.

19.27 For a particle moving in a magnetic field the frequency of its motion is

$$f = \frac{1}{T} = \frac{qB}{2\pi m} = \frac{(1.60 \times 10^{-19}\ C)(1.50\ T)}{2\pi(3.34 \times 10^{-27}\ kg)} = 1.14 \times 10^7\ Hz,$$

$f = \boxed{11.4\ MHz}$.

19.29 For a particle moving in a magnetic field the frequency of its motion is

$$f = \frac{1}{T} = \frac{qB}{2\pi m} = \frac{(1.60 \times 10^{-19}\ C)(1.0\ T)}{2\pi(9.1 \times 10^{-31}\ kg)} = 2.8 \times 10^{10}\ Hz = \boxed{28\ GHz}.$$

19.31 From Eq. (19.4), $mv = qBr$.

$$KE = \frac{(mv)^2}{2m} = \frac{q^2B^2r^2}{2m}.$$

19.33 The magnetic field of a long wire is given by Eq. (19.5)

$B = k'\frac{2I}{d}$, where $k' = 1.00 \times 10^{-7}\ N/A^2$.

$$B = 1.00 \times 10^{-7}\ N/A^2 \times \frac{2(100 \times 10^{-3}\ A)}{0.100\ m} = \boxed{2.00 \times 10^{-7}\ T}.$$

19.35 (a) $\frac{F}{l} = IB = k'\frac{2I^2}{d} = (1.00 \times 10^{-7}\ N/A^2)\left(\frac{2(200\ A)^2}{0.50\ m}\right) = \boxed{0.016\ N/m}$.

(b) The wires $\boxed{\text{repel}}$ each other.

19.37 $\frac{F}{l} = I_2B_1 = \frac{k'2I_1I_2}{d}$.

$$I_2 = \frac{(F/l)d}{2k'I_1} = \frac{(1.00 \times 10^{-3}\ N/m)(2.50 \times 10^{-2}\ m)}{(2 \times 10^{-7}\ N/A^2)(100\ A)} = \boxed{1.25\ A}.$$

19.39 The field of each loop has magnitude $\frac{\mu_o I}{2r}$ and is directed along the axis of the loop. The resulting field is the vector sum of the individual contributions from each loop, thus the sum is N times the field due to one loop.

$$B = N\frac{\mu_o I}{2r} = \frac{\mu_o NI}{2r}.$$

19.41 Currents in the same direction attract, currents in the opposite direction repel. Let us denote the attractive force as F_+ and the repulsive force as F_-.

$$F_+ = \frac{2k'I_1I_2l}{d_+} \text{ and } F_- = \frac{2k'I_1I_2l}{d_-}.$$

$$F_{net} = F_+ - F_- = 2k'I_1I_2l\left(\frac{1}{d_+} - \frac{1}{d_-}\right),$$

$$F_{net} = 2(10^{-7}\ \text{N/A}^2)(6.3\ \text{A})(5.4\ \text{A})(0.40\ \text{m})\left(\frac{1}{0.02\ \text{m}} - \frac{1}{0.12\ \text{m}}\right),$$

$$F_{net} = \boxed{1.1 \times 10^{-4}\ \text{N attractive}}.$$

Note that for the sides of the loop that are perpendicular to the long wire, their currents interact with the field of the long wire to produce equal but opposite forces. Thus the net contribution from these sides is zero.

19.43 Area of loop is $A = \pi r^2 = \pi(d/2)^2$. $d/2 = 0.065$ m.

$$\mu = 150\ (0.075\ \text{A})(\pi)(0.0625\ \text{m})^2 = \boxed{0.14\ \text{A}\cdot\text{m}^2}.$$

19.45 $\tau = \mu B \sin\theta = NIAB \sin\theta$,

$$\tau = (6)(0.200\ \text{A})(0.043 \times 0.058\ \text{m}^2)(0.055\ \text{T})(1/\sqrt{2}) = \boxed{1.2 \times 10^{-4}\ \text{N}\cdot\text{m}}.$$

19.47 Use Ohm's law to find the voltage on the meter in terms of the current through the meter and the total resistance.

$$V_m = IR_m = I(R_s + R_G) = (100\ \mu\text{A})(10\ \text{k}\Omega + 490\ \text{k}\Omega),$$

$$V_m = (100 \times 10^{-6}\ \text{A})(500 \times 10^3\ \Omega) = \boxed{50\ \text{V}}.$$

19.49 (a) First find the resistance of the voltmeter circuit that draws 10.0 mA when a voltage of 200 V is applied.

$$R_m = 200\ \text{V}/0.0100\ \text{A} = 20\,000\ \Omega.$$

The voltmeter resistance is the sum of the meter resistance R_G and a series resistance R_s. The series resistance is

$$R_s = R_m - R_G = \boxed{19\,000\ \Omega\ \text{in series}}.$$

(b) The circuit has the form shown.

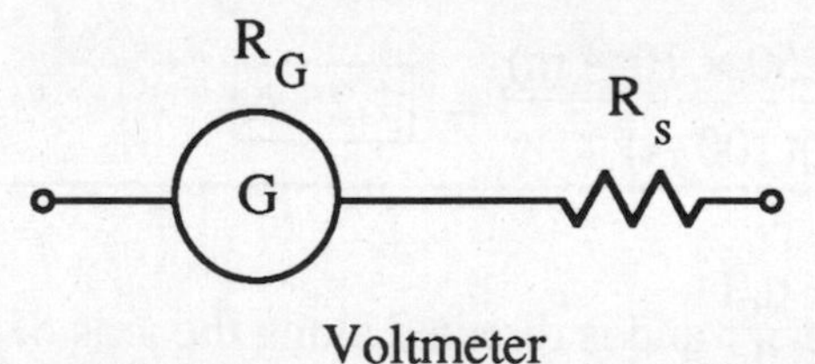

Voltmeter

19.51 $$B = \mu_o nI = (4\pi \times 10^{-7}\ \text{N/A}^2)\left(\frac{900}{0.435\ \text{m}}\right)(2.75\ \text{A}) = \boxed{7.15\ \text{mT}}.$$

19.53 The current can be obtained from Eq. (19.12).

$$I = \frac{B}{\mu_o n} = \frac{(0.800\ \text{T})(2.00\ \text{m})}{(4\pi \times 10^{-7}\ \text{N/A}^2)(320)} = \boxed{4000\ \text{A}}.$$

19.55 Consider a circular path centered about the axis of the conductor at a distance $r < a$ from the center of the conductor. Applying Ampere's law we get:

$\sum B\,\Delta l = \mu_o I_{enclosed}$. Because of the symmetry the field must be the same magnitude along the path so that

$$\sum B\,\Delta l = B \sum \Delta l = B(2\pi r).$$

The current enclosed is given by: $\dfrac{\text{area enclosed}}{\text{total area}} I = \dfrac{r^2}{a^2} I$. So

$$B(2\pi r) = \mu_o \frac{r^2}{a^2} I, \qquad B = \boxed{\frac{\mu_o I r}{2\pi a^2}}.$$

19.57 Rearranging Eq. (19.4) we find the electric charge to be

$$q = \frac{mv}{rB} = \frac{2.03 \times 10^{-22}\ \text{kg·m/s}}{(0.0823\ \text{m})(7.69 \times 10^{-5}\ \text{T})} = \boxed{3.21 \times 10^{-19}\ \text{C}} = 2e.$$

19.59 Combine Eq. (19.4) with the expression for the field of a solenoid:

$$B = \frac{mv}{qr} = \mu_o n I.$$

$$n = \frac{mv}{\mu_o I r q} = \frac{(1.67 \times 10^{-27}\ \text{kg})(2.00 \times 10^6\ \text{m/s})}{(4\pi \times 10^{-7}\ \text{N/A}^2)(1000\ \text{A})(2.00\ \text{m})(1.60 \times 10^{-19}\ \text{C})},$$

$$n = \boxed{8.31\ \text{turns/m}}.$$

19.61 The total length L is fixed. If the wire is coiled into N loops, the circumference of each loop is $l_n = 2\pi r = L/N$. The magnetic moment is $\mu = NIA$.

$$\mu = NI\pi r^2 = NI\pi\left(\frac{L}{2\pi N}\right)^2 = \frac{IL^2}{4\pi N}.$$

$\boxed{\text{The maximum moment occurs for } N = 1}$.

19.63 The velocity component perpendicular to the field can be obtained from Eq. (19.4) as

$$v_\perp = \frac{rqB}{m} = \frac{(0.0512\ \text{m})(1.60 \times 10^{-19}\ \text{C})(5.70 \times 10^{-2}\ \text{T})}{3.34 \times 10^{-27}\ \text{kg}} = 1.40 \times 10^5\ \text{m/s}.$$

The pitch is given by $v_{||}T$, where T is the time to complete one rotation about the axis and is given by $T = \dfrac{2\pi r}{v_\perp}$. Thus the parallel component of the velocity is

$$v_{||} = \frac{\text{pitch}}{T} = \frac{(\text{pitch})(v_\perp)}{2\pi r} = \frac{(0.167\ \text{m})(1.40 \times 10^5\ \text{m/s})}{2\pi(0.0512\ \text{m})} = 7.26 \times 10^4\ \text{m/s}.$$

The speed of the deuteron is

$$v = \sqrt{(v_\perp)^2 + (v_{||})^2} = \boxed{1.58 \times 10^5\ \text{m/s}}.$$

Chapter 20

ELECTROMAGNETIC INDUCTION

20.1 For the situation depicted in the drawing in the text, the field is greatest and the change in the field is greatest when the pole of the magnet is nearest the center of the coil as shown below. Consequently, the voltage output for a uniformly rotating magnet is as shown. You can test this for yourself with a coil and a magnet.

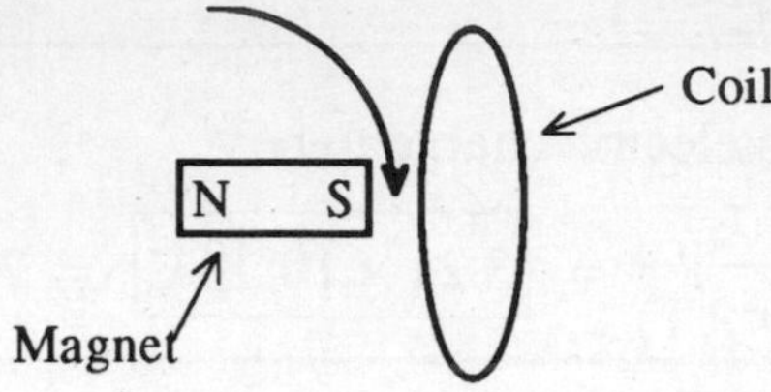

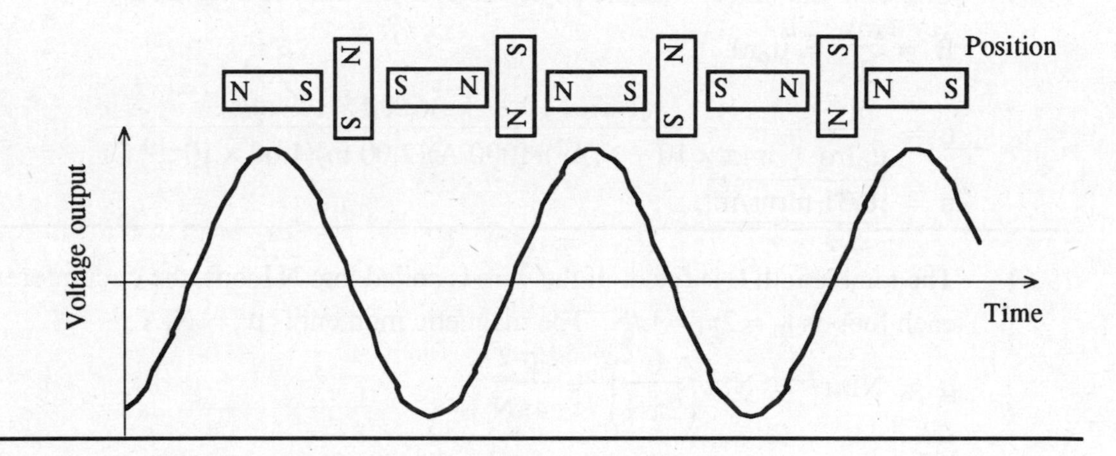

20.3 $$|\mathscr{E}| = N\frac{\Delta\phi}{\Delta t} = \frac{NA\,\Delta B}{\Delta t},$$

$$|\mathscr{E}| = \frac{(1)(37 \times 10^{-4}\ m^2)(9.3 - 6.5)(10^{-3}\ T)}{0.50\ s} = \boxed{21\ \mu V}.$$

20.5 The emf is related to the magnetic field through Eq. (20.3).

$$|\mathscr{E}| = N\frac{\Delta\phi}{\Delta t} = \frac{NA\Delta B}{\Delta t}.$$

Here $\Delta t = t = 0.25$ s and $\Delta B = B - 0$. Consequently,

$|\mathscr{E}| = \frac{NAB}{t}$. Upon rearranging we find that

$$B = \frac{\mathscr{E}t}{NA} = \frac{(0.30 \times 10^{-3}\ V)(0.25\ s)}{(300)(25 \times 10^{-4}\ m^2)} = \boxed{1.0 \times 10^{-4}\ T}.$$

20.7 (a) $|\mathcal{E}| = N\dfrac{\Delta\phi}{\Delta t} = \dfrac{NAB}{t} = \dfrac{(25)(10 \times 10^{-4}\ m^2)(0.20\ T)}{0.50\ s} = \boxed{10\ mV}$.

(b) $I = \dfrac{\mathcal{E}}{R} = \dfrac{10\ mV}{75\ \Omega} = \boxed{0.13\ mA}$.

20.9 (a) From Eq. (20.3) we get the magnitude of the emf to be

$|\mathcal{E}| = \dfrac{NA\Delta B}{\Delta t}$, where $\Delta B = B/2$. The area of the coil is $A = \pi d^2/4$ where d is the coil diameter. Consequently,

$$|\mathcal{E}| = \frac{N(\pi d^2/4)(B/2)}{\Delta t} = \frac{(3\ \pi)(0.17\ m)^2(3.8 \times 10^{-2}\ T)}{8(0.27\ s)} = \boxed{4.8\ mV}.$$

(b) Now $\Delta B = 2B$, so $|\mathcal{E}| = \dfrac{N(\pi d^2/4)(2B)}{\Delta t} = \boxed{19\ mV}$.

(c) Now $\Delta B = B$, so $|\mathcal{E}| = \boxed{9.6\ mV}$.

(d) Now $\Delta B = 2B$, so $|\mathcal{E}| = \boxed{19\ mV}$.

20.11 $\mathcal{E} = Blv = (1.75 \times 10^{-5}\ T)(47.65\ m)\left(\dfrac{8.5 \times 10^5\ m}{3600\ s}\right) = \boxed{0.20\ V}$.

20.13 $\mathcal{E} = Blv$. Just before striking the ground, $v = \sqrt{2gh}$.

$\mathcal{E} = (4.0 \times 10^{-5}\ T)(1.0\ m)\left(\sqrt{2(9.8\ m/s^2)(12\ m)}\right) = \boxed{0.61\ mV}$.

20.15 For a single loop the peak emf is $\mathcal{E}_{peak} = \omega BA$. For a loop of N turns the emf is $\mathcal{E}_{peak} = N\omega BA$. The angular frequency $\omega = 2\pi f$.

$\mathcal{E}_{peak} = N2\pi fBA = (50)(2\pi)(10\ Hz)(5 \times 10^{-5}\ T)(0.010\ m^2) = \boxed{1.6\ mV}$.

20.17 $\mathcal{E} = N2\pi f\, BA\sin\theta = (1)(2\pi)(5.0\ Hz)(1.3 \times 10^{-2}\ T)(0.0120\ m^2)\sin 0.27°$,

$\mathcal{E} = \boxed{2.3 \times 10^{-5}\ V}$.

20.19 From Eq. (20.6) we find

$V_2 = \dfrac{N_2}{N_1}V_1 = \dfrac{50}{500}\,120\ V = \boxed{12\ V}$.

20.21 From Eq. (20.6) we find

$\dfrac{N_1}{N_2} = \dfrac{V_1}{V_2} = \dfrac{120\ V}{12\ V} = \boxed{10}$.

20.23 From Eq. (20.7) we have $V_1I_1 = V_2I_2$. Rearranging to get I_1 we find

$I_1 = \dfrac{V_2}{V_1}I_2 = \left(\dfrac{10\ V}{120\ V}\right)(1.5\ A) = \boxed{0.125\ A}$.

20.25 The output voltage V_2 of transformer T_1 is the input voltage V_3 of transformer T_2. So $V_3 = V_2 = \frac{N_2}{N_1} V_1$,

$$V_4 = \frac{N_4}{N_3} V_3 = \frac{N_4 N_2}{N_3 N_1} V_1.$$

20.27 The input power P_1 must equal the output power P_2. $P_1 = V_1 I_1 = V_2 I_2 = P_2$.

So $I_1 = \left(\frac{V_2}{V_1}\right) I_2$.

From Eq. (20.6) $V_2/V_1 = N_2/N_1$. Also $I_2 = V_2/R$, so that

$$I_1 = \left(\frac{N_2}{N_1}\right)\frac{V_2}{R} = \left(\frac{N_2}{N_1}\right)\left(\frac{N_2}{N_1}\right)\frac{V_1}{R} = \left(\frac{N_2}{N_1}\right)^2 \frac{V_1}{R}.$$

20.29 Find the power loss.

$P_{loss} = P_{in} - P_{out} = (3750\text{ V})(20\text{ A}) - (1200\text{ V})(60.25\text{ A}) = 2700\text{ W}.$

$P = \frac{Q}{t} = \left(\frac{m}{t}\right) c\Delta T$, where c is the specific heat.

$$\Delta T = \frac{P}{(m/t)c} = \frac{(2700\text{ W})(60\text{ s})}{(5.0\text{ kg})(3000\text{ J/kg°C})} = 10.8\text{°C}.$$

$$T_{out} = T_{in} + \Delta T = 15\text{°C} + 10.8\text{°C} = \boxed{25.8\text{°C}}.$$

20.31 $$\mathcal{E} = -L\frac{\Delta I}{\Delta t} = -(0.80\text{ H})(-0.10\text{ A/s}) = \boxed{80\text{ mV}}.$$

20.33 This is the same as Example 20.7.

$$L = \frac{\mu_o N^2 A}{\ell},$$

$$N = \sqrt{\frac{\ell L}{\mu_o A}} = \sqrt{\frac{(0.050\text{ m})(1.3\times 10^{-3}\text{ H})}{(4\pi\times 10^{-7}\text{ H/m})(12\times 10^{-4}\text{ m}^2)}} = \boxed{208}.$$

20.35 $\mu_o = 4\pi\times 10^{-7}\text{ N/A}^2$. Examine the unit combination N/A^2.

$$\frac{\text{N}}{\text{A}^2} = \frac{\text{J/m}}{\text{A}^2} = \frac{\text{J/m}}{\text{A}\cdot\text{C/s}} = \frac{\text{J}\cdot\text{s}}{\text{m}\cdot\text{C}\cdot\text{A}} = \frac{(\text{V}\cdot\text{s})/\text{A}}{\text{m}} = \frac{\text{H}}{\text{m}}.$$

20.37 From Problem 20.36 the inductance is given by $L = \frac{\mu_o N^2 A}{l}$.

$$L = \frac{(4\pi\times 10^{-7}\text{ N/A}^2)(200)^2(6.0\times 10^{-4}\text{ m}^2)}{0.30\text{ m}} = \boxed{100\ \mu\text{H}}.$$

20.39 From Eq. (20.11) we get $W = \frac{1}{2}LI^2$. Rearranging,

$$I = \sqrt{\frac{2W}{L}} = \sqrt{\frac{2(4.0\times 10^{-5}\text{ J})}{38\times 10^{-3}\text{ H}}} = \boxed{46\text{ mA}}.$$

20.41 $c = \frac{1}{\sqrt{\mu_o \varepsilon_o}} = \frac{1}{\sqrt{(4\pi \times 10^{-7}\ N/A^2)(8.854 \times 10^{-12}\ C/V\cdot m)}}$,

$c = \boxed{2.998 \times 10^8\ m/s}$.

20.43 If the force is smaller, then the effective value of the constant in Coulomb's law is smaller, thus "k" $= \frac{k}{4000} \Rightarrow$ "ε_o" $= 4000\ \varepsilon_o$.

"c" $= \frac{c}{\sqrt{4000}} = \boxed{4.740 \times 10^6\ m/s}$.

20.45 (a) $emf = \frac{\Delta\phi}{\Delta t} = \frac{B\Delta A}{\Delta t} = Bl\frac{\Delta x}{\Delta t} = \boxed{Blv}$.

(b) $emf = \boxed{Blv}$.

(c) $I = \frac{\mathcal{E}}{R} = \boxed{\frac{Blv}{R}}$.

(d) $P = VI = Fv$.

$F = \frac{(Blv)^2}{Rv} = \boxed{\frac{B^2l^2v}{R}}$.

20.47 (a) $P = \frac{V^2}{R}$. The potential drop across the power lines is the difference between the potential at the generator and the potential at the other end. So

$P = \frac{[(13{,}800 - 12000)V]^2}{6.2\ \Omega} = \boxed{520\ kW}$.

(b) New power loss to be 10^{-3} of the old power, or

$P = I^2R = 10^{-3}\ I_o^2R$, where I and I_o are the new and old currents.

$I = \frac{I_o}{\sqrt{1000}}$.

$P_{transformer\ in} = P_{transformer\ out}$, or $VI = V_oI_o$,

$V = V_o\frac{I}{I_o} = V_o\sqrt{1000} = (13{,}800\ V)(\sqrt{1000}) = \boxed{440\ kV}$.

(c) Cost $= (P_o - P)$ (t)(rate).

$P_o = 520$ kW, $P = 0.52$ kW, so that $P_o - P \approx P_o$.

Cost $=$ (520 kW)(365 d)(24 h/d)($0.08/kWh) $=$ $\boxed{\$360{,}000}$.

20.49 The energy stored in the field is $W = \frac{1}{2}LI^2$.

The energy density is $u = \frac{W}{V}$, where V is the volume occupied by the field.

$u = \frac{\frac{1}{2}LI^2}{lA} = \frac{\frac{1}{2}(\mu_oN^2A/l)I^2}{lA} = \frac{1}{2\mu_o}\left(\frac{\mu_oNI}{l}\right)^2 = \frac{B^2}{2\mu_o}$.

20.51 $\mathcal{E} = Blv.$

$$P = \mathcal{E}I = \frac{\mathcal{E}^2}{R} = \frac{B^2l^2v^2}{R} = \frac{B^2l^2x^2}{Rt^2}, \text{ where we used } v = x/t.$$

$$W = Pt = \frac{B^2l^2x^2}{Rt} = \frac{(0.15\ T)^2(0.10\ m)^2(1.0\ m)^2}{(200\ \Omega)(0.25\ s)} = \boxed{4.5\ \mu J}.$$

20.53 $\mathcal{E} = A\frac{\Delta B}{\Delta t}$, where $B = \mu_o nI$.

$$\mathcal{E} = A_{loop}\,\mu_o n\frac{\Delta I}{\Delta t},$$

$$W = Pt = \frac{\mathcal{E}^2}{R}t = \frac{A^2\mu_o{}^2n^2}{R}\left(\frac{\Delta I}{\Delta t}\right)^2 t,$$

$$W = \frac{\pi^2(0.025\ m)^4(\mu_o{}^2)(2.0\times 10^4)^2}{1.6\ \Omega}(0.020\ A/s)^2\ (3.0\ s),$$

$$W = \boxed{1.8\times 10^{-12}\ J}.$$

20.55 We need $V_A - V_C = 0 = IR - L\frac{\Delta I}{\Delta t}$. $\boxed{\text{Yes}}$.

20.57 For the parallel connection, the potential difference across the two inductors must be the same, so $\mathcal{E} = -L_1\frac{\Delta I_1}{\Delta t} = -L_2\frac{\Delta I_2}{\Delta t}$.

Total current is $I = I_1 + I_2$. The rate of change of the current is thus

$$\frac{\Delta I}{\Delta t} = \frac{\Delta I_1}{\Delta t} + \frac{\Delta I_2}{\Delta t}.$$

But $\frac{\Delta I_1}{\Delta t} = -\frac{\mathcal{E}}{L_1}$, and $\frac{\Delta I_2}{\Delta t} = -\frac{\mathcal{E}}{L_2}$.

$\frac{\Delta I}{\Delta t} = -\mathcal{E}\left(\frac{1}{L_1}+\frac{1}{L_2}\right) = -\frac{\mathcal{E}}{L_p}$, if we define the effective parallel inductance as

$$\frac{1}{L_p} = \left(\frac{1}{L_1}+\frac{1}{L_2}\right)$$

20.59 From Problem 20.56, $L = L_1 + L_2$.

$L_2 = L - L_1 = 0.87\ H - 0.27\ H = \boxed{0.60\ H}$.

20.61 To get the maximum inductance we $\boxed{\text{joining inductors in series}}$ to get

$L = 3\,L_1 = \boxed{0.75\ H}$.

Chapter 21

ALTERNATING-CURRENT CIRCUITS

21.1 From Eq. (20.8), the defining equation for inductance, $LI = N\phi_m$, we see that the unit of inductance is

$[L] = \frac{[\phi_m]}{[I]}$. The unit for magnetic flux is $[\phi_m] = [B][\text{area}] = T{\cdot}m^2$.

$H = \frac{T{\cdot}m^2}{A}$. The unit for magnetic field, the tesla, is $T = \frac{N}{A \cdot m}$.

We find the units of henries/ohm by substituting these units.

$$\frac{H}{\Omega} = \frac{T{\cdot}m^2}{A \cdot \Omega} = \frac{\frac{N}{A \cdot m} m^2}{A \cdot \Omega} = \frac{N \cdot m}{A^2 \cdot \Omega} = \frac{J}{A^2(V/A)} = \frac{J \cdot s}{C \cdot V} = s.$$

21.3 From the definition of the time constant, $\tau_L = \frac{L}{R}$.

$$R = \frac{L}{\tau_L} = \frac{0.17\ H}{0.13\ s} = \boxed{1.3\ \Omega}.$$

21.5 Equation (21.1) gives the current as a function of time: $I = \frac{V_o}{R}\left(1 - e^{-t/\tau}\right)$,

Substitute in the values for current, battery voltage, and resistance to get

$80\ mA = \frac{9.0\ V}{100\ \Omega}\left(1 - e^{-Rt/L}\right)$. Next rearrange to get

$\left(1 - e^{-Rt/L}\right) = \frac{(0.080\ A)(100\ \Omega)}{9.0\ V} = \frac{8}{9}$. Again rearrangc to find

$e^{-Rt/L} = 1 - \frac{8}{9} = \frac{1}{9}$. Take the log of both sides to get

$$\frac{-Rt}{L} = -\ln 9,$$

$$t = \frac{L}{R}\ln 9 = \frac{50 \times 10^{-3}\ H}{100\ \Omega}(2.20) = \boxed{1.1\ ms}.$$

21.7 (a) $\tau_L = \frac{L}{R} = \frac{0.80\ H}{5.0\ \Omega} = \boxed{0.16\ s}$.

(b) $\frac{V}{V_o} = 0.99 = 1 - e^{-t/\tau}$,

$e^{-t/\tau} = 1 - 0.99 = 0.01$. Now take the logarithm of both sides to get

$-t/\tau_L = \ln(0.01)$,

$t = -\tau_L \ln(0.01) = (-0.16\ s)(-4.6) = \boxed{0.74\ s}$.

21.9 $\tau_C = RC = (3.9 \times 10^6\ \Omega)(250 \times 10^{-6}\ F) = \boxed{980\ s}$.

21.11 The parallel capacitors add to give an effective value of $C = 10{,}000\ \mu F$. The resistance is $R = 110\ k\Omega$.

(a) $\tau_C = RC = (1.1 \times 10^5\ \Omega)(1.0 \times 10^{-2}\ F) = \boxed{1100\ s}$.

(b) $V_C = V_o e^{-t/\tau} = (50\ V)\, e^{-3600/1100} = \boxed{1.9\ V}$.

21.13 (a) For discharging a capacitor the voltage divided by the initial voltage is

$$\frac{V}{V_o} = \frac{10}{1500} = e^{-t/\tau},$$

$e^{t/\tau} = 150$. Now take the logarithms to find $t = \tau_C \ln(150) = \boxed{5\tau_C}$.

(b) $t = 5\tau = 5RC = (5)(10^5\ \Omega)(0.47 \times 10^{-6}\ F) = \boxed{0.24\ s}$.

21.15 $$f = \frac{1}{T} = \frac{1}{20.0 \times 10^{-3}\ s} = \boxed{50.0\ Hz}.$$

21.17 $$I_{rms} = \frac{I_m}{\sqrt{2}} = \frac{17\ mA}{\sqrt{2}} = \boxed{12\ mA}.$$

21.19 $$V_{rms} = \sqrt{\overline{v^2}} = \sqrt{\left(V_o^2\right)\left(\overline{\sin^2 2\pi ft}\right)} = V_o\sqrt{\frac{1}{2}} = \boxed{\frac{V_o}{\sqrt{2}}}.$$

21.21 (a) $$X_C = \frac{1}{2\pi fC} = \frac{1}{2\pi(60\ s^{-1})(1.0 \times 10^{-6}\ F)} = \boxed{2.7\ k\Omega},$$

(b) $$X_C = \frac{1}{2\pi(500\ s^{-1})(1.0 \times 10^{-6}\ F)} = \boxed{320\ \Omega},$$

(c) $$X_C = \frac{1}{2\pi(1000\ s^{-1})(1.0 \times 10^{-6}\ F)} = \boxed{160\ \Omega},$$

(d) $$X_C = \frac{1}{2\pi(10000\ s^{-1})(1.0 \times 10^{-6}\ F)} = \boxed{16\ \Omega}.$$

21.23 $$f = \frac{1}{2\pi C X_C} = \frac{1}{2\pi(10^{-4}\ F)(4.0\ \Omega)} = \boxed{400\ Hz}.$$

21.25 $X_L = 2\pi fL = 2\pi(2050\ Hz)(525 \times 10^{-3}\ H) = \boxed{6.76\ k\Omega}$.

21.27 $X_L = 2\pi fL$. $$f = \frac{X_L}{2\pi L} = \frac{100\ \Omega}{2\pi(25 \times 10^{-3}\ H)} = \boxed{640\ Hz}.$$

21.29 (a) From Eq.(21.18) we see that

$$I_C = \frac{V_{rms}}{X_C} = 2\pi fCV_{rms} = 2\pi(60\ s^{-1})(1.5 \times 10^{-6}\ F)(80\ V) = \boxed{45\ mA}.$$

(b) The maximum value of the current is

$I_m = \sqrt{2}\, I_{rms} = \sqrt{2}\,(45\text{ mA}) = \boxed{64\text{ mA}}$.

21.31 The current can be found from

$$I_m = \frac{V_m}{Z} = \frac{V_m}{\sqrt{R^2 + (X_L - X_C)^2}} = \frac{V_m}{\sqrt{R^2 + \left(2\pi fL - \dfrac{1}{2\pi fC}\right)^2}}.$$

$$I_m = \frac{10}{\sqrt{50^2 + \left((2\pi)(2000)(0.010) - \dfrac{1}{(2\pi)(2000)(10^{-6})}\right)^2}} = \boxed{0.15\text{ A}}.$$

21.33 First find the impedance of the circuit: $Z = \sqrt{R^2 + (2\pi fL)^2}$.

(a) $I = \dfrac{V}{Z} = \dfrac{120\text{ V}}{\sqrt{(330)^2 + [(2\pi)(60)(0.80)]^2}\ \Omega} = \boxed{0.27\text{ A}}$.

(b) $\phi = \tan^{-1}\dfrac{X_L}{R} = \boxed{42°}$.

(c) $P = I^2R = (0.27\text{ A})^2(330\ \Omega) = \boxed{24\text{ W}}$.

21.35 (a) $\phi = \tan^{-1}\dfrac{-X_C}{R} = \tan^{-1}\dfrac{-1}{2\pi fCR} = \tan^{-1}\dfrac{-1}{(2\pi)(100)(10^{-4})(330)}$,

$\phi = \boxed{-2.8°}$.

(b) Phase angle between the capacitor voltage and the resistor voltage is $\boxed{-90°}$.

(c) The current has the same phase through the circuit so the phase angle is $\boxed{0}$.

21.37 $f_o = \dfrac{1}{2\pi\sqrt{LC}} = \dfrac{1}{2\pi\sqrt{(2.5\times10^{-3}\text{ H})(0.010\times10^{-6}\text{ F})}} = \boxed{32\text{ kHz}}$.

21.39 (a) $f_o = \dfrac{1}{2\pi\sqrt{LC}} = \dfrac{1}{2\pi\sqrt{(0.350\text{ H})(0.15\times10^{-6}\text{ F})}} = \boxed{690\text{ Hz}}$.

(b) At $f = f_o$, $Z = R = \boxed{1500\ \Omega}$.

21.41 At $f = f_o$, the impedance $Z = R = 51\ \Omega$.

$i = \dfrac{V_m}{R}\cos 2\pi ft = \dfrac{12\text{ V}}{51\ \Omega}\cos 2\pi ft = (0.24\text{ A})\cos 2\pi ft$.

The frequency f_o is found from

$2\pi f_o = 1/\sqrt{LC} = 1/\sqrt{(10\times10^{-3}\text{ H})(1.0\times10^{-6}\text{ F})} = 10000\text{ Hz}$.

Thus the current becomes $i = \boxed{(0.24\text{ A})\cos(10000\ t)}$.

21.43 (a) The energy stored in the capacitor is $\frac{1}{2}CV^2$. At some point all of the energy is transferred to the inductor where it can be expressed as $\frac{1}{2}LI^2$, where I is the maximum current. We can find the value of the current maximum by equating these two expressions.

$$\frac{1}{2}LI^2 = \frac{1}{2}CV^2.$$

$$I = V\sqrt{\frac{C}{L}} = (30\ V)\sqrt{\frac{10 \times 10^{-6}\ F}{0.45\ H}} = \boxed{0.14\ A}.$$

(b) $$f = \frac{1}{2\pi\sqrt{LC}} = \frac{1}{2\pi\sqrt{(0.45\ H)(10 \times 10^{-6}\ F)}} = \boxed{75\ Hz}.$$

21.45 $$I = \frac{V_o}{R}\left(1 - e^{-t/\tau}\right).$$

(a) When $t = \frac{\tau_L}{2}$, $I = \frac{V_o}{R}\left(1 - e^{-1/2}\right) = \boxed{0.39\frac{V_o}{R}}$.

(b) When $t = 2\tau_L$, $I = \frac{V_o}{R}\left(1 - e^{-2}\right) = \boxed{0.86\frac{V_o}{R}}$.

(c) When $t = 4\tau_L$, $I = \frac{V_o}{R}\left(1 - e^{-4}\right) = \boxed{0.98\frac{V_o}{R}}$.

21.47 (a) $\tau_C = RC = (500 \times 10^3\ \Omega)(1000 \times 10^{-6}\ F) = \boxed{500\ s}$.

(b) $V = V_o e^{-t/\tau} = (100\ V)\ e^{-120/500} = \boxed{79\ V}$.

21.49 (a) The impedance is computed from Eq.(21.21b).

$$Z = \sqrt{R^2 + \left(2\pi fL - \frac{1}{2\pi fC}\right)^2},$$

$$Z = \sqrt{(470)^2 + \left((2\pi)(365)(0.225) - \frac{1}{(2\pi)(365)(1.50 \times 10^{-6})}\right)^2} = \boxed{521\ \Omega}.$$

(b) $$f_o = \frac{1}{2\pi\sqrt{LC}} = \frac{1}{2\pi\sqrt{(0.225\ H)(1.50 \times 10^{-6}\ F)}} = \boxed{274\ Hz}.$$

(c) $Z = \boxed{470\ \Omega}$.

21.51 The capacitor voltage builds up according to $V_C = V_o\left(1 - e^{-t/RC}\right)$.
Upon rearranging we find

$$e^{-t/RC} = 1 - \frac{V_C}{V_o} = 1 - \frac{10}{12} = \frac{1}{6}.$$

Take the logarithm of both sides to get

$$-t/RC = -\ln 6,$$

$$t = RC \ln 6 = (1000\ \Omega)(500 \times 10^{-6}\ F)\ 1.79 = \boxed{0.90\ s}.$$

Chapter 22

GEOMETRICAL OPTICS

22.1 Use a diagram similar to the one in Example 22.1.

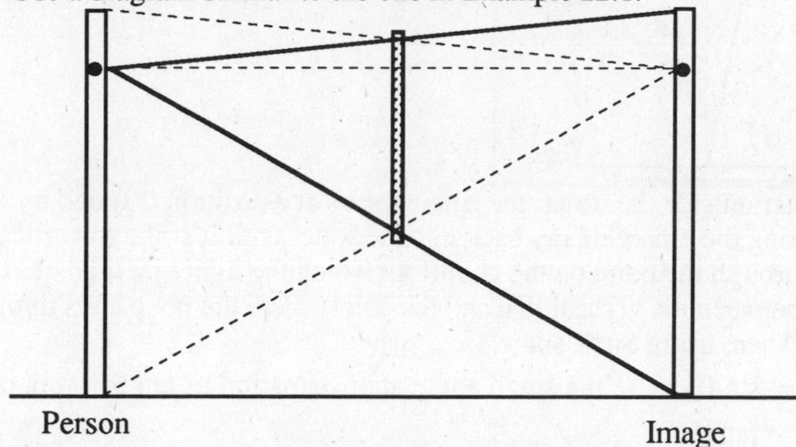

By analyzing the geometry we find the height needed for the mirror to be $h' = (152 \text{ cm})/2 = \boxed{76 \text{ cm}}$.

22.3 From the geometry of the situation and from the law of reflection we find that the reflected light ray will emerge $\boxed{\text{parallel to the symmetry axis}}$ moving in the direction opposite to the incoming light.

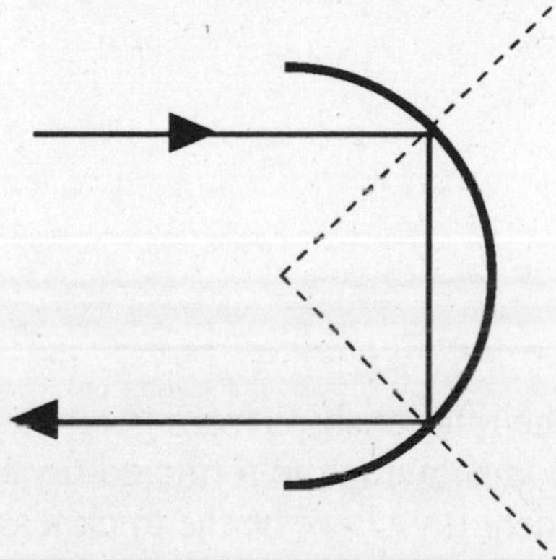

22.5 For a plane mirror, the distance between the mirror and your image is the same as the distance between the mirror and yourself. If you move a distance d toward the mirror, your image also moves a distance d toward the mirror. If you run toward the mirror at a speed of 2.7 m/s, the image will also approach the mirror at a speed 2.7 m/s. Consequently, you will approach you image at twice that rate or $\boxed{5.4 \text{ m/s}}$.

22.7 A ray of light leaving the coin at depth h strikes the surface of the water at an angle θ_1 and emerges into the air at an angle θ_2, where the angles are related by Snell's law with $n_1 = n$ and $n_2 = 1$.

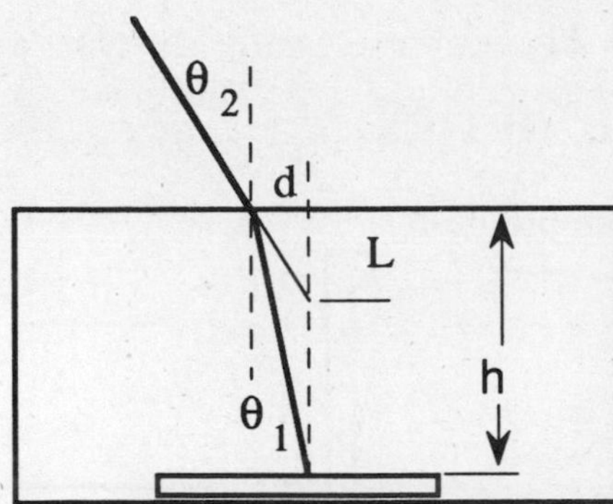

To a viewer outside the water, the coin appears at a position L found by extrapolating the emerging ray back into the water until it strikes a vertical line passing through the point on the coin from which the light emerged. Let d be the distance between the vertical line and the point where the ray passes through the surface. Then, using Snell's law,

$n \sin \theta_1 = \sin \theta_2$, Use the small angle approximation of $\tan \theta \approx \sin \theta$,

$n \tan \theta_1 = \tan \theta_2$,

$n \frac{d}{h} = \frac{d}{L}$, or $L = \frac{h}{n} = \frac{38 \text{ cm}}{1.33} = \boxed{29 \text{ cm}}$.

22.9 **Three** images can be seen.

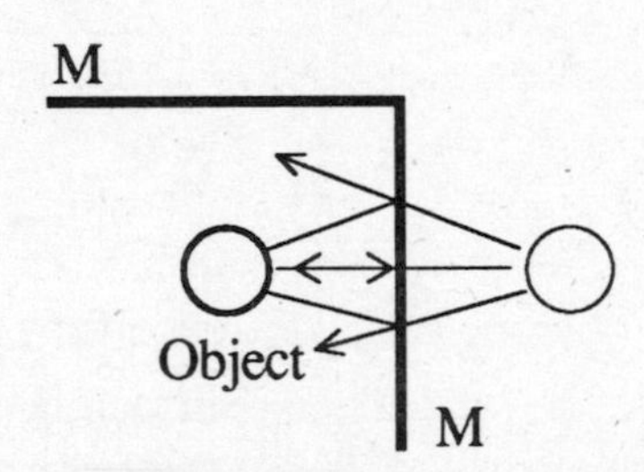

22.11 (a) You can show this graphically using the relationships from geometry.
(b) Draw a diagram of the physical situation showing the refracted ray and extrapolating a line for the undeviated incident ray. Let t be the thickness of the glass block and L be the length of the refracted ray in the glass.

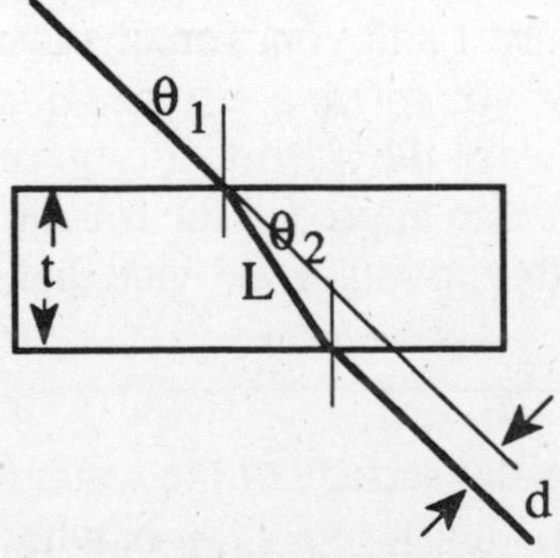

From the geometry, the length L is seen to be $L = \frac{t}{\cos \theta_2}$, where θ_2 is the angle of the refracted beam inside the glass. Now imagine a line extending from the

point of contact of the refracted ray with the second surface of the glass and striking the extrapolated incident ray path perpendicularly. This line is the displacement d of the transmitted ray from the path of the incident ray. $d = L \sin(\theta_1 - \theta_2)$. So the displacement is $d = \boxed{\dfrac{t\sin(\theta_1 - \theta_2)}{\cos\theta_2}}$.

22.13 $\theta_c = \sin^{-1}\left(\frac{n_2}{n_1}\right) = \sin^{-1}\left(\frac{1.000}{2.417}\right) = \boxed{24.44°}$.

22.15 Angular diameter $= 2\,\theta_c = 2\sin^{-1}\left(\frac{1.000}{1.333}\right) = \boxed{97.2°}$.

22.17 Make a sketch of the situation.

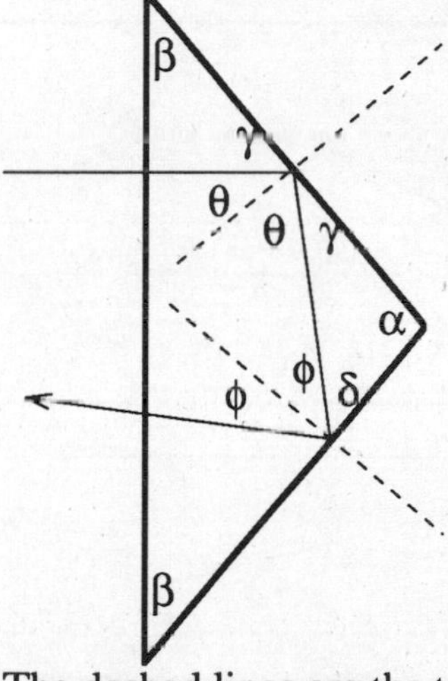

The dashed lines are the two surface normals.The angle of incidence for the first rear surface is θ, and the angle of incidence for the second rear surface is ϕ. Because the incident light is normal to the first surface, you can see from the diagram that $\beta + \gamma = 90°$. Also we have $2\beta + \alpha = 180°$, and $\theta + \gamma = 90°$. From this you can get $\theta = \beta$ and that $\alpha + 2\theta = 180°$.
We can express the last equation as
$\alpha = 180° - 2\theta$.
In order for the light to be internally reflected at the first reflection, the angle of incidence θ must be greater than or equal to the critical angle θ_c. That is $\theta \geq \theta_c$.
Thus we get $\alpha = 180° - 2\theta \leq 180° - 2\theta_c$.
We can compute the critical angle from $\theta_c = \sin^{-1}\frac{1}{1.55} = 40.2°$. This gives an upper limit for α of $\alpha \leq 180° - 2(40.2°) = 99.6°$.
Upper limit is α is 99.6°.
The angle of incidence at the second rear surface is $\phi = 90° - \delta$, where $\alpha + \delta + \gamma = 180°$. But we have already seen that $\theta = \beta$ so that $\gamma = 90° - \beta$, so we can rewrite the expression for $\alpha + \delta + \gamma$ as

$\alpha + \delta + (90° - \beta) = 180°$, or $\alpha - \beta = 90° - \delta$

Then $\phi = (90° - \delta) = \alpha - \beta \geq \theta_c = 40.2°$.

But we know that $\alpha = 180° - 2\theta = 180° - 2\beta$. So

$\beta = 90° - \alpha/2$.

$\alpha - \beta = \alpha - (90° - \alpha/2) = \frac{3}{2}\alpha - 90° \geq 40.2°$. Thus we get

$\frac{3}{2}\alpha \geq 90° + 40.2°$.

So, $\frac{3}{2}\alpha \geq 130.2°$. Upon rearranging we get

$\alpha \geq 86.8°$ This gives a lower limit for α of $86.8°$.

For total reflection, the values of α must lie between 86.8° and 99.6°.

$\boxed{86.8° \leq \alpha \leq 99.6°}$.

22.19 First make a careful scale drawing.

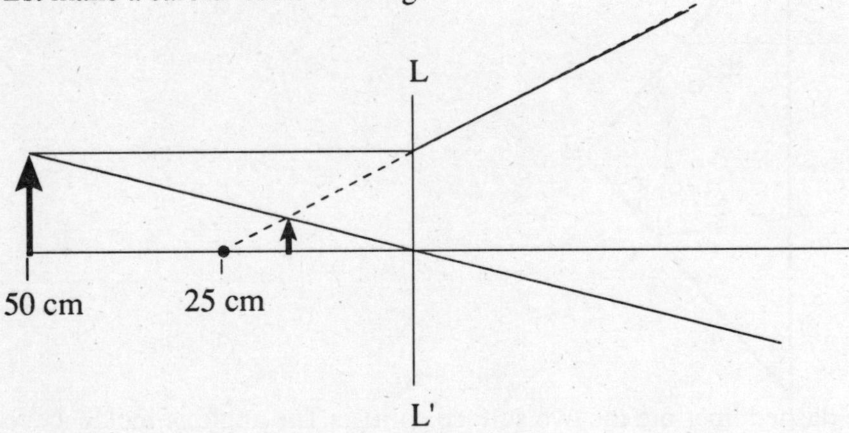

(a) From the drawing, the lens-to-stamp distance is $\boxed{17\text{ cm}}$.

(b) The magnification is $m = \boxed{+0.33}$.

22.21 Make a scale drawing.

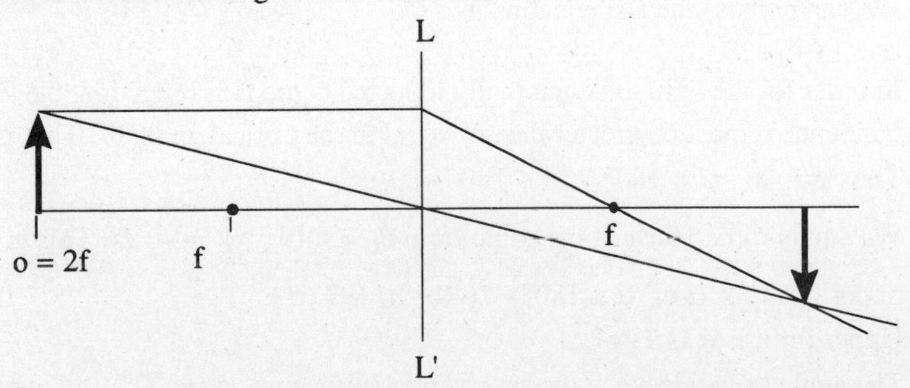

From measurement of the drawing we see that the image occurs at a distance 2f from the lens.

22.23 Use the ray-tracing procedure of Section 22.5.

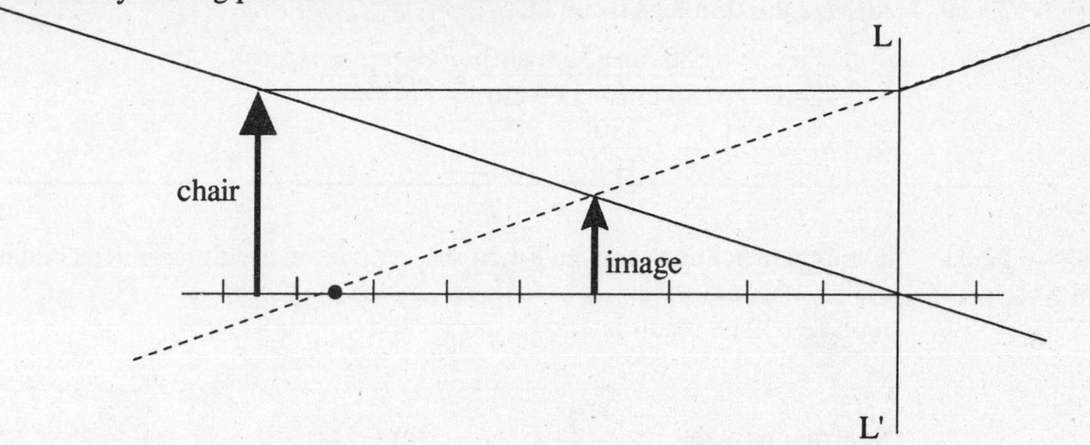

Remember that the focal length of the diverging lens is negative. A ray from the chair is undeviated when passing through the center of the lens. A ray from the chair parallel to the axis emerges through the focal point. The chair is 0.86 m away from the lens.

22.25 Draw the position of the bulb and screen to scale on an optical axis. Represent the bulb by an upright arrow and its image by an inverted arrow three times as large.
(a) Draw a ray from the tip of the object arrow to the tip of the image arrow.

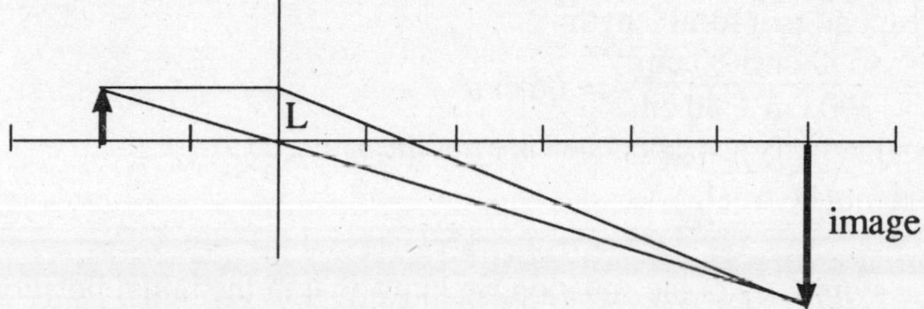

This ray passes through the center of the lens at the point where it crosses the optic axis, at the 30 cm mark.
(b) Draw a ray parallel to the optic axis from the tip of the object arrow to a line marking the position of the lens. Draw another line from this intersection to the tip of the image arrow. This ray crosses the optic axis at the focal point of the lens, 15 cm from the lens.

22.27 Use the thin lens equation and let o = 2f. $i = \frac{of}{o - f} = \frac{(2f)(f)}{2f - f} = 2f.$
The object and images distances are the same, so the magnification is 1. Note that this is true only for a converging lens.

22.29 (a) Use the thin lens formula: $\frac{1}{o} + \frac{1}{i} = \frac{1}{f}$,

$o = \frac{if}{i-f} = \frac{(180\text{ cm})(35.6\text{ cm})}{180\text{ cm} - 35.6\text{ cm}} = \boxed{44.4\text{ cm}}$.

(b) $m = -\frac{i}{o} = -\frac{180}{44.4} = \boxed{-4.1\times}$.

22.31 The magnification is given by $m = -\frac{i}{o}$. Now use the thin lens equation to replace o. The result is $m = -\frac{i}{\frac{if}{i-f}} = -\frac{i-f}{f}$.

Rearrange to get $fm = f - i$, $f(m-1) = -i$, $f = \frac{i}{1-m}$.

Here $i = 1.27$ m and $m = -2.5$. (A real, inverted image has a negative m.)

$f = \frac{1.27\text{ m}}{1-(-2.5)} = \boxed{0.36\text{ m}}$.

22.33 First calculate the focal length of the lens.

$f = \frac{oi}{o+i} = \frac{(250\text{ cm})(50\text{ cm})}{250\text{ cm} + 50\text{ cm}} = 42\text{ cm}$.

For the new positions, the distance between object and lens is 390 cm and between lens and screen is 80 cm. If the new claimed values were true, they would correspond to a focal length

$f = \frac{oi}{o+i} = \frac{(390\text{ cm})(80\text{ cm})}{390\text{ cm} + 80\text{ cm}} = 66\text{ cm}$.

The two focal lengths just calculated are not the same, so

$\boxed{\text{the claim cannot be true}}$.

22.35 (a) From the symmetry of the situation we know that in the initial position the lens must be 15 cm from the bulb. From the thin lens equation, $\frac{1}{f} = \frac{1}{o} + \frac{1}{i}$, we get $f = \frac{oi}{o+i}$. Here $o = 15$ cm and $i = 160$ cm. Solving for f we find

$f = \frac{(15\text{ cm})(160\text{ cm})}{175\text{ cm}} = \boxed{13.7\text{ cm}}$.

(b) The flashlight bulb is $i + o = \boxed{175\text{ cm}}$ from the wall.

22.37 The longitudinal magnification M_L is the magnification along the optic axis.

$M_L = \frac{\Delta i}{\Delta o}$. From the thin lens equation we get

$f = \frac{o_1 i_1}{o_1 + i_1} = \frac{o_2 i_2}{o_2 + i_2}$, where the subscripts refer to two different object and image points. Eliminating the fractions gives

$o_1 i_1(o_2 + i_2) = o_2 i_2(o_1 + i_1)$,

$o_1 o_2 i_1 + o_1 i_1 i_2 = o_1 o_2 i_2 + o_2 i_1 i_2$,

$o_1 o_2 i_1 - o_1 o_2 i_2 = -o_1 i_1 i_2 + o_2 i_1 i_2$,

$$o_1o_2(i_1 - i_2) = - i_1i_2(o_1 - o_2),$$

$$M_L = \frac{i_1 - i_2}{o_1 - o_2} = -\frac{i_1i_2}{o_1o_2} = -m^2.$$

22.39 For the mirror, $f = R/2 = 15$ cm.

$$i = \frac{of}{o-f} = \frac{(20\text{ cm})(15\text{ cm})}{20\text{ cm} - 15\text{ cm}} = \boxed{60\text{ cm}} \text{ on the side with the object.}$$

$$h' = mh = -\frac{i}{o}h = -\frac{60\text{ cm}}{20\text{ cm}}2.0\text{ cm} = \boxed{-6.0\text{ cm}}.$$

22.41 (a) $f = \dfrac{oi}{o+i} = \dfrac{(87\text{ cm})(18\text{ cm})}{87\text{ cm} + 18\text{ cm}} = \boxed{15\text{ cm}}$.

(b) $R = 2f = \boxed{30\text{ cm}}$.

(c) $m = -\dfrac{i}{o} = -\dfrac{18}{87} = \boxed{-0.21}$.

22.43 (a) $f = R/2 = -95$ mm/4.

$$i = \frac{of}{o-f} = \frac{(2000\text{ mm})(-95\text{ mm}/4)}{2024\text{ mm}} = \boxed{-23.5\text{ mm}}.$$

(b) $m = -\dfrac{i}{o} = -\dfrac{-23.5\text{ mm}}{2000\text{ mm}} = \boxed{0.012}$.

(c) The image is $\boxed{\text{erect}}$.

22.45 Because the ratio of image distance to object distance is equal to the ratio of the image height to object height (the magnification), the image distance must be 2 times the object distance. You can only see such an image of yourself if the image is virtual and thus lies behind the mirror. Thus for positive o, the i is negative.

$m = -\dfrac{i}{o} = 2.$ So, $i = -2o$. Using the mirror equation we get

$$\frac{1}{o} - \frac{1}{2o} = \frac{1}{f} = \frac{2}{R}.$$

$\dfrac{1}{2o} = \dfrac{2}{R}$, so the object distance is $o = \boxed{R/4}$.

22.47 Use Snell's law with $n_1 = 1$ and $n_2 = n$. $n = \dfrac{\sin\theta_1}{\sin\theta_2} = \dfrac{\sin 47°}{\sin 32.5°} = 1.36.$

From the table we see that the material is probably $\boxed{\text{ethyl alcohol}}$.

22.49 (a) $\theta_c = \sin^{-1}\left(\dfrac{1.00}{1.31}\right) = 49.8°$. $\boxed{\text{No}}$, the light will not be totally reflected because $\theta_i < \theta_c$.

(b) The symmetric ice prism that is closest to a 45°–45°–90° prism that will totally reflect an incident normal beam is $\boxed{50°\text{–}50°\text{–}80°\text{ prism}}$.

22.51 From Fig. 22.19 we see that

$\frac{f}{h_i} = \frac{x}{h_o}$ and that $\frac{f}{h_o} = \frac{x'}{h_i}$. We can rearrange to get

$f = \frac{h_o}{h_i}x' = \frac{x}{f}x'$ or $f^2 = xx'$.

22.53 First consider the thin lens equation for each lens.

$\frac{1}{o_1} + \frac{1}{i_1} = \frac{1}{f_1}$, and $\frac{1}{o_2} + \frac{1}{i_2} = \frac{1}{f_2}$.

When the two lenses are very close together we can make the approximation that $o_2 = -i_1$. Substitute this into the first equation to get

$$\frac{1}{o_1} + \frac{1}{i_1} = \frac{1}{o_1} + \frac{1}{-o_2} = \frac{1}{o_1} + \left(\frac{1}{i_2} - \frac{1}{f_2}\right) = \frac{1}{f_1}.$$

Now rearrange the final two terms to get

$$\frac{1}{o_1} + \frac{1}{i_2} = \frac{1}{f_1} + \frac{1}{f_2}.$$

This is of the form $\frac{1}{o} + \frac{1}{i} = \frac{1}{f}$, where $\frac{1}{f} = \frac{1}{f_1} + \frac{1}{f_2}$.

22.55 Since the object is very far away, consider the incident light to be a parallel beam. The incident light is imaged by the lens to a point 16 cm away which is the focal point of the mirror. After passing through the focal point the light strikes the mirror and is reflect back as a beam parallel to the axis. When the reflected light again strikes the lens it will be focused to a point 16 cm from the lens. Thus the final image of the distant object is located **16 cm in front of the lens**.

22.57 The speed of light is $c = \frac{2D}{t}$, where t is the time for the toothed wheel to turn from one gap to the next. For a wheel with N teeth, it requires a time $t = \frac{2\pi}{N\omega}$ for the wheel to rotate from one gap to the next. Thus the speed of light can be found from $\boxed{c = \frac{ND\omega}{\pi}}$.

22.59 For the first lens,

$$i_1 = \frac{o_1 f_1}{o_1 - f_1} = \frac{(18\text{ cm})(12\text{ cm})}{18\text{ cm} - 12\text{ cm}} = 36\text{ cm}.$$

This image becomes the object for the second lens. Since the second lens is 18 cm beyond the first lens, the new object position is $o_2 = 18\text{ cm} - i_1 = -18\text{ cm}$. The second image is located at

$$i_2 = \frac{o_2 f_2}{o_2 - f_2} = \frac{(-18\text{ cm})(-6.0\text{ cm})}{-18\text{ cm} - (-6.0\text{ cm})} = \frac{18 \times 6\text{ cm}}{-12\text{ cm}} = -9\text{ cm}.$$

Assuming light propagating from left to right, the final image is located **9 cm to the left of the diverging lens**. This device is a Galilean telescope.

22.61 Assume propagation from left to right. The image formed by the first lens is located at $i_1 = \dfrac{o_1 f_1}{o_1 - f_1} = \dfrac{(36\text{ cm})(24\text{cm})}{36\text{ cm} - 24\text{ cm}} = 72$ cm.

The image by the first lens becomes the object for the second lens at an object distance $o_2 = 18\text{ cm} - 72\text{ cm} = -54$ cm.

$$i_2 = \frac{o_2 f_2}{o_2 - f_2} = \frac{(-54\text{ cm})(-18.0\text{ cm})}{-54\text{ cm} - (-18.0\text{ cm})} = \frac{54 \times 18\text{ cm}}{-36\text{ cm}} = -27\text{ cm.}$$

The object position for the third lens is $o_3 = 9\text{ cm} - (-27\text{ cm}) = 36$ cm.

$$i_3 = \frac{o_3 f_3}{o_3 - f_3} = \frac{(36\text{ cm})(12\text{ cm})}{36\text{ cm} - 12\text{ cm}} = \frac{36 \times 12\text{ cm}}{24\text{ cm}} = 18\text{ cm.}$$

The final image is **18 cm to the right of the third lens**.

Chapter 23

OPTICAL INSTRUMENTS

23.1 $f = \frac{1}{\text{strength}} = \frac{1}{\mathscr{D}} = \frac{1}{+5.00\text{ D}} = 0.200\text{ m} = \boxed{20.0\text{ cm}}$.

23.3 From Problem 22.53 we are given that $\frac{1}{f} = \frac{1}{f_1} + \frac{1}{f_2}$. This is equivalent to saying

$\mathscr{D} = \mathscr{D}_2 + \mathscr{D}_1 = (+5) + (-2) = +3$ diopters.

$f = \frac{1}{\mathscr{D}} = \boxed{0.33\text{ m}}$.

23.5 Use the thin lens equation with $o = \infty$ and $i = -1.00$ m.

$$\frac{1}{f} = \frac{1}{o} + \frac{1}{i} = \frac{1}{\infty} + \frac{1}{-1.00\text{ m}} = \frac{1}{-1.00\text{ m}},$$

$f = -1.00$ m. Strength $= \frac{1}{f} = \boxed{-1.00\text{ diopter}}$.

23.7 Assume the glasses correct to the standard near point of 25 cm.

$$\frac{1}{f} = +2.00/\text{m} = \frac{1}{o} + \frac{1}{i} = \frac{1}{0.25\text{ m}} + \frac{1}{i},$$

$$\frac{1}{i} = 2.00/\text{m} - \frac{1}{0.25\text{ m}} = -2.00/\text{m},$$

$i = -0.50$ m. The person's near point is $\boxed{0.50\text{ m}}$.

23.9 First find the old far point, which is the same as the focal length of the lens.

far point $= \frac{1\text{ m}}{-2.70}$. If the far point moves inward by 20%, then it becomes

$\frac{0.8\text{ m}}{-2.70}$ and the strength of the lens required is $\frac{-2.70}{0.8\text{ m}} = \boxed{-3.4\text{ diopters}}$.

23.11 From Eq. 23.4 we get $f = \frac{25\text{ cm}}{M} = \frac{25\text{ cm}}{4} = \boxed{6.3\text{ cm}}$.

23.13 $\theta' = M\theta = M\frac{d}{R} = 5\,\frac{1.8\text{ cm}}{25\text{ cm}} = \boxed{0.36\text{ rad}}$.

23.15 (a) $\frac{1}{i} = \frac{1}{f} - \frac{1}{o} = \frac{M}{25\text{ cm}} - \frac{1}{4.0\text{ cm}} = \frac{8}{25\text{ cm}} - \frac{1}{4.0\text{ cm}},$

$i = \boxed{14\text{ cm}}$.

(b) The image is $\boxed{\text{real}}$.

23.17 Separately we get $M = \frac{25}{f} = \frac{25}{10} = \boxed{2.5\times}$ and $M = \frac{25}{16} = \boxed{1.6\times}$.

In combination we get the sum, $M = \boxed{4.1\times}$.

23.19 Make a diagram similar to Fig. 23.4 with the lens spaced a distance d from the eye. Label the height of the image h′ and the distance from the eye to the image as $s = i + d$, where both i and d are magnitudes.

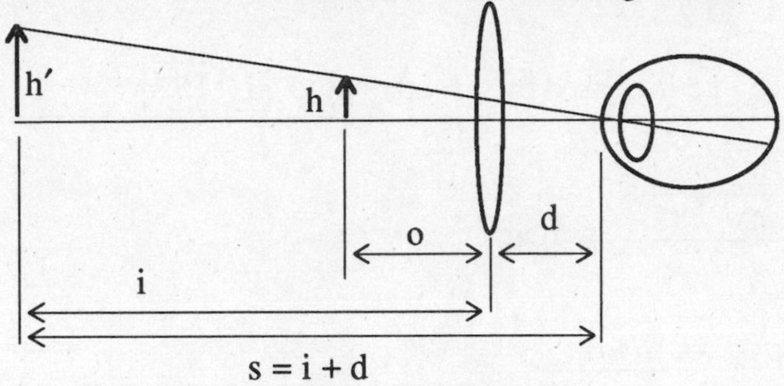

Then the angle becomes

$\theta' = \frac{h'}{s}$. The magnification of a thin lens is $\frac{h'}{h} = \frac{i}{o}$, so that $\theta' = \frac{h}{s}\frac{i}{o}$.

The angle θ, corresponding to the unaided eye is $\theta = \frac{h}{25}$ with h in centimeters.

The magnification required is

$$M = \frac{\theta'}{\theta} = \frac{25}{h}\left(\frac{h}{s}\frac{i}{o}\right)$$

We can use the thin lens equation to locate the object position:

$\frac{1}{o} + \frac{1}{-i} = \frac{1}{f}$, where we have explicitly shown that the image distance is negative. Thus $\frac{1}{o} = \frac{i+f}{if}$. Then use $i = s - d$ to get

$$M = \frac{25}{s}\left(\frac{s-d+f}{f}\right) = \frac{25}{f} + \frac{25}{s} - \frac{25d}{sf}.$$

23.21 In going from f/1.4 to f/2.8, lens diameter decreases by 2, so the light passed decreases by 4. The shutter speed (time) must be increased by 4 to maintain the same exposure. Set the shutter for $\boxed{1/125 \text{ s}}$.

23.23 $\text{f-number} = \frac{f}{d} = \frac{17 \text{ m}}{5 \text{ m}} = \boxed{f/3.4}$.

23.25 Focal length = (f-number)(diameter) = (2.8)(8 mm) = $\boxed{2.2 \text{ cm}}$.

22.27 The maximum f-number is $\frac{55 \text{ mm}}{15 \text{ mm}} = f/3.7$. If the shutter speed is increased by a factor of 10, the f-number must be decreased by a factor of $\sqrt{10} = 3.16$. Thus the initial f-number of f/5.6 needs to become

$$\text{f-number} = \frac{f/5.6}{3.16} = f/1.8.$$

This value is beyond the limiting value of f/3.7 found above.

$\boxed{\text{No, the shutter cannot be set for 1/500 s.}}$

The amount of light is increased by

$$\left(\frac{f/old}{f/new}\right)^2 = \left(\frac{5.6}{3.7}\right)^2 = 2.3.$$

The shutter speed can be increased by that amount to $\boxed{1/115 \text{ s}}$.

23.29 $$\frac{1}{(\text{f-number})^2} = \frac{\frac{1}{(1.4)^2} + \frac{1}{(2)^2}}{2} = 0.38,$$

$$\text{f-number} = \sqrt{\frac{1}{0.38}} = \boxed{f/1.6}.$$

23.31 Consider the triangle formed by the optic axis, a ray through the center of the lens extending to the corner of the film, and the film plane.

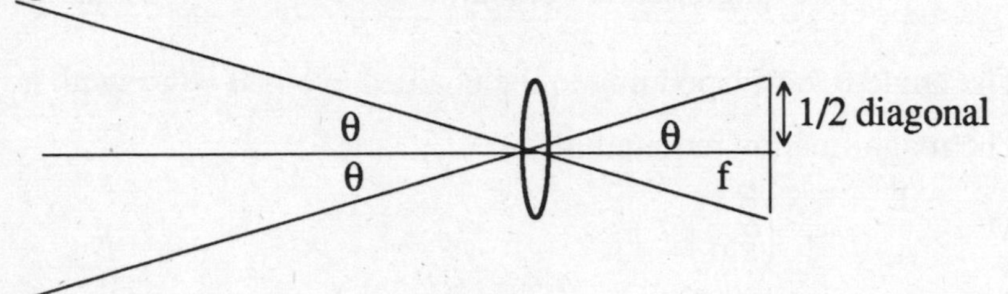

Call the angle between the optic axis and the ray θ. The angle of view of the camera is

$2\theta = 2\tan^{-1}\frac{\text{film diagonal}/2}{f} = 2\tan^{-1}\frac{f}{2f} = 2\tan^{-1}\frac{1}{2}$. Notice that the angle of view is independent of the focal length of the camera. Hence, for a "normal" lens, the field of view is independent of camera format.

23.33 First find the distance from the lens to the slide (assuming thin lens equivalent).

$$o = \frac{if}{i-f} = \frac{(427 \text{ cm})(17.9 \text{ cm})}{427 \text{ cm} - 17.9 \text{ cm}} = 18.7 \text{ cm}.$$

The linear magnification is $m = h_i/h_o = i/o$. Thus the image height on the screen is $h_i = \frac{h_o i}{o} = \frac{(23.5 \text{ mm})(427 \text{ cm})}{18.7 \text{ cm}} = 537 \text{ mm} = 0.537 \text{ m}$.

The width on the screen is $h_i = \frac{(34.3 \text{ mm})(427 \text{ cm})}{18.7 \text{ cm}} = 783 \text{ mm} = 0.783 \text{ m}$.

The screen must have at least the dimensions of $\boxed{0.537 \text{ m} \times 0.783 \text{ m}}$.

23.35 First find the magnification of the objective.

$$|m_o| = \frac{2000\times}{20\times} = 100\times.$$

Then determine the focal length from the relation that

$$f = \frac{L}{|m_o|} = \frac{160 \text{ mm}}{100\times} = \boxed{1.6 \text{ mm}}.$$

23.37 (a) $|m_o| = \frac{L}{f_o} = \frac{160 \text{ mm}}{3.2 \text{ mm}} = \boxed{50\times}$.

$M_e = \frac{25 \text{ cm}}{f_e} = \frac{25 \text{ cm}}{2.5 \text{ cm}} = \boxed{10\times}$.

(b) $M = m_o M_e = (50\times)(10\times) = \boxed{500\times}$.

23.39 For a standard microscope the magnification is $M_s = \left(\frac{16.0 \text{ cm}}{f_o}\right)(M_e)$.

For the 16.5 cm microscope, the magnification is $M = \left(\frac{16.5 \text{ cm}}{f_o}\right)(M_e)$. The ratio of magnifications is $\frac{M}{M_s} = \frac{16.5 \text{ cm}}{16.0 \text{ cm}} = \boxed{1.03}$.

23.41 $M = -\frac{f_o}{f_e} = -\frac{50 \text{ cm}}{3.0 \text{ cm}} = \boxed{-17\times}$.

23.43 The magnification of the eyepiece is given by $M_e = \frac{25 \text{ cm}}{f_e}$. Rearrange to get

$f_e = \frac{25 \text{ cm}}{M_e} = \frac{25 \text{ cm}}{5\times} = 5 \text{ cm}$.

For a very distant object the distance between objective and eyepiece is $f_e + f_o$.

$f_o = 55 \text{ cm} - 5 \text{ cm} = 50 \text{ cm}$.

$M = -\frac{f_o}{f_e} = -\frac{50}{5} = \boxed{-10\times}$.

23.45 $\theta' = M\theta = 7 \times \frac{1.90 \text{ m}}{50 \text{ m}} = \boxed{0.27 \text{ rad}}$.

23.47 (a) f_e − total separation f_o = $\boxed{20 \text{ cm}}$.

(b) $M = -\frac{f_o}{f_e} = -\frac{50 \text{ cm}}{-20 \text{ cm}} = \boxed{+2.5\times}$.

23.49 Separation between the lenses should be the image distance of the first lens plus the (negative) focal length of the eyepiece. The image distance is

$i = \frac{of}{o-f} = \frac{(30 \text{ m})(20 \text{ cm})}{29.8 \text{ m}} = 20.1 \text{ cm}$.

If the eyepiece forms an image at infinity, the separation is 20.1 cm – 5.0 cm = $\boxed{15.1 \text{ cm}}$.

23.51 (a) A distant object is imaged by the converging lens at its focal length. This image becomes the virtual object for the diverging lens with an object distance of $o = -5$ cm. The image formed by the diverging lens is at

$i = \frac{of}{o-f} = \frac{(-5 \text{ cm})(-10 \text{ cm})}{-5 \text{ cm} + 10 \text{ cm}} = \boxed{+10 \text{ cm}}$.

(b) When the separation between the lenses is 8 cm, o = – 2 cm, i becomes

$$i = \frac{(-2\text{ cm})(-10\text{ cm})}{-2\text{ cm} + 10\text{ cm}} = \boxed{+2.5\text{ cm}}.$$

23.53 Compare f-numbers. For the 10-cm focal length lens,

$$\text{f-number} = \frac{f}{d} = \frac{10\text{ cm}}{4.0\text{ cm}} = f/2.5.$$

For the 24-cm lens, f-number $= \frac{f}{d} = \frac{24\text{ cm}}{6.0\text{ cm}} = f/4.0$.

The $\boxed{\text{smaller diameter lens}}$ will produce the brightest image.

23.55 The sun is so far away that the light comes to focus at the focal point of the lens.

$$M = \frac{25\text{ cm}}{f} = \frac{25\text{ cm}}{15\text{ cm}} = \boxed{1.7\times}.$$

23.57 (a) In pairs, the combinations are $\boxed{50\times, 100\times, 200\times, 400\times}$.

(b) Again in pairs, $\boxed{100\times, 150\times, 200\times, 300\times, 400\times, 600\times}$.

23.59 In Fig. 23.9(a) there is a triangle formed by the incident ray that passes through the center of the objective lens, the optic axis, and the line that represents the objective image. There is a second triangle with a common side formed by the ray through the center of the eyepiece, the optic axis, and the objective image. From examination of the diagram and the definition of an angle, the angular magnification is found to be $M = \frac{\theta'}{\theta} = -\frac{f_o}{f_e}$.

23.61 In new position i = 31 cm. We know f = 30 cm, so find o using the thin lens equation.

$$o = \frac{if}{i-f} = \frac{(31\text{ cm})(30\text{ cm})}{31\text{ cm} - 30\text{ cm}} = 930\text{ cm} = \boxed{9.3\text{ m}}.$$

23.63 (a)

$$\text{N.A.} = \sin\alpha/2 = \frac{(0.92\text{ cm})/2}{\sqrt{\left(\frac{0.92\text{ cm}}{2}\right)^2 + (1.79\text{ cm})^2}} = \boxed{0.25}.$$

(b)

$$\text{f-number} = \frac{f}{d} = \frac{1.79\text{ cm}}{0.92\text{ cm}} = \boxed{1.95}.$$

(c) The light entering the objective is approximately proportional to the area of the aperture which is proportional to $(\sin\alpha/2)^2 = (\text{N.A.})^2$.

Chapter 24

WAVE OPTICS

24.1 Geometric construction of expanding waves.

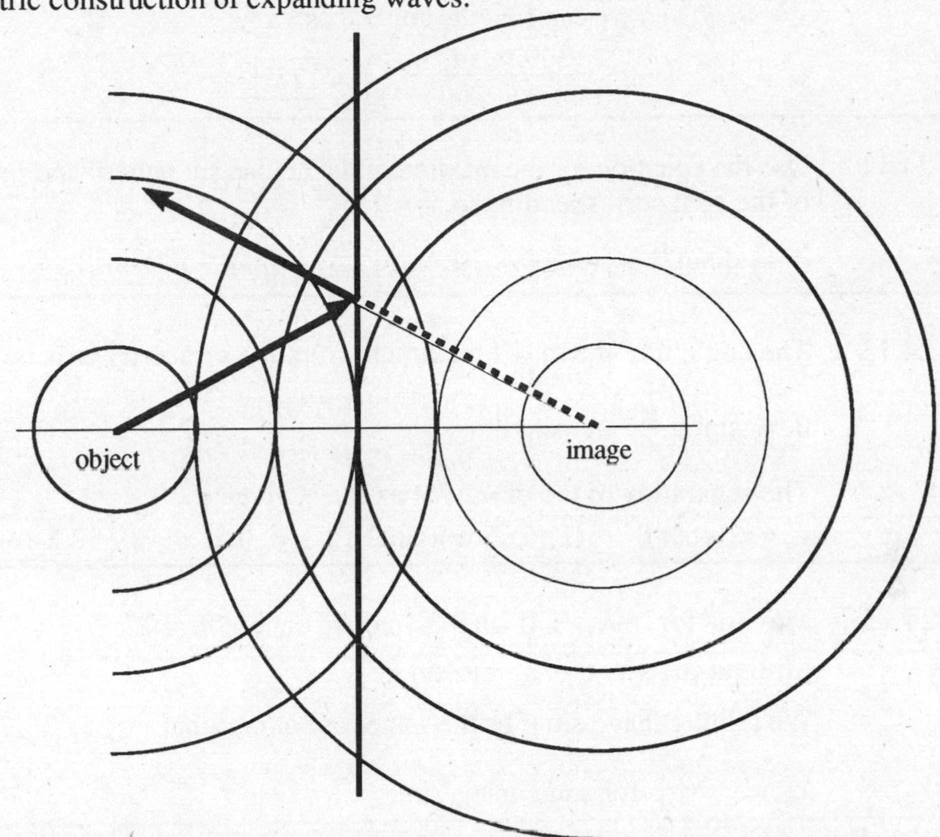

24.3 (a) The wavelength λ' in water is less then the wavelength λ in air by a factor of one over the index of refraction. $\lambda' = \lambda/n = (589 \text{ nm})/1.333 = \boxed{442 \text{ nm}}$.

(b) $\lambda' = \lambda/n = (589 \text{ nm})/1.49 = \boxed{395 \text{ nm}}$.

24.5 (a) Speed of light in the glass is

$v = c/n = (3.00 \times 10^8 \text{ m/s})/1.65 = \boxed{1.82 \times 10^8 \text{ m/s}}$.

(b) $\lambda' = \lambda/n = (633 \text{ nm})/1.65 = \boxed{384 \text{ nm}}$.

24.7 Use the equation for the double slit maximum with m = 2, λ = 550 nm and d = 0.50 mm. $m\lambda = d \sin \theta$,

$$\theta = \sin^{-1}\left(\frac{m\lambda}{d}\right) = \sin^{-1}\left(\frac{(2)(550 \times 10^{-9} \text{ m})}{0.50 \times 10^{-3} \text{ m}}\right) = \boxed{0.13°}.$$

24.9 First order minimum occurs when $\frac{1}{2}\lambda = d \sin\theta$. Solve for θ.

$\theta = \sin^{-1}\frac{\lambda}{2d} = \sin^{-1}\frac{v}{2df}$, where we use the relationship that $v = f\lambda$. Then we can find the angle θ as

$$\theta = \sin^{-1}\left(\frac{340 \text{ m/s}}{(2)(0.20 \text{ m})(2000 \text{ s}^{-1})}\right) = \boxed{25°}.$$

24.11 Use the equation for the maxima of the double slit pattern and look for the angle of the spot corresponding to m = 1.

$$\lambda = d \sin\theta = (0.500 \times 10^{-3} \text{ m})\frac{0.106 \text{ cm}}{90.0 \text{ cm}} = \boxed{589 \text{ nm}}.$$

24.13 The angle θ of the m = 7 maximum from the symmetry axis is

$$\theta = \sin^{-1}\left(\frac{7\lambda}{d}\right) = \sin^{-1}\left(\frac{(7)(633 \times 10^{-9} \text{ m})}{0.120 \times 10^{-3} \text{ m}}\right) = 2.12°.$$

The separation of the $m = \pm 7$ maxima is given by

$$y = 2L \tan\theta = (2)(5.75 \text{ m})(\tan 2.12°) = 0.51 \text{ m} = \boxed{42.5 \text{ cm}}.$$

24.15 Maxima for $m\lambda_1 = d \sin\theta$. Here we choose m = 1.

Minima for $(m + \frac{1}{2})\lambda_2 = d \sin\theta$.

We require that $d \sin\theta$ be the same for both, so that

$$\lambda_2 = \frac{\lambda_1}{m + \frac{1}{2}} \text{ for some integer m.}$$

For m = 0, $\lambda_2 = 2\lambda_1 = 2(440 \text{ nm}) = 880$ nm, in the infra red.

For m = 1, $\lambda_2 = \frac{2}{3}\lambda_1 = \frac{2}{2}(440 \text{ nm}) = 293$ nm, in the ultraviolet.

$\boxed{\text{No wavelength of visible light is possible.}}$

24.17 From the equation for the double slit maxima we get

$\frac{\lambda_1}{d} = \sin\theta \approx \frac{y_1}{L}$, for light with wavelength λ_1 and $\frac{\lambda_2}{d} \approx \frac{y_2}{L}$, for the light with wavelength λ_2. L and d are constant. So $\frac{L}{d} = \frac{\lambda_1}{y_1} = \frac{\lambda_2}{y_2}$.

$$y_2 = y_1\frac{\lambda_2}{\lambda_1} = (1.00 \text{ mm})\frac{656 \text{ nm}}{434 \text{ nm}} = \boxed{1.51 \text{ mm}}.$$

24.19 For minimum reflection we want $2t = m\frac{\lambda}{n}$, where m is 0, 1, 2, ... and n is the index of refraction of the mica (n = 1.58). Minimum thickness for m = 1.

$$t = \frac{\lambda}{2n} = \frac{450 \text{ nm}}{(2)(1.58)} = \boxed{142 \text{ nm}}.$$

24.21 For maximum reflection, $t = \frac{\lambda}{2n} = \frac{560\text{ nm}}{(2)(1.32)} = \boxed{212\text{ nm}}$.

24.23 For maximum reflection we want $2t = \left(m + \frac{1}{2}\right)\frac{\lambda}{n}$, where m is 0, 1, 2, ...

For $m = 0$, $\lambda = 4\,nt = 4\,(1.50)(400\text{ nm}) = 2400\text{ nm}$. (Not visible.)

For $m = 1$, $\lambda = \frac{4}{3}nt = \frac{4}{3}(1.50)(400\text{ nm}) = 800\text{ nm}$. (Not visible.)

For $m = 2$, $\lambda = \frac{4}{5}nt = \frac{4}{5}(1.50)(400\text{ nm}) = 480\text{ nm}$. (Visible.)

For $m = 2$, $\lambda = \frac{4}{7}nt = \frac{4}{7}(1.50)(400\text{ nm}) = 343\text{ nm}$. (Not visible.)

The best reflected visible wavelength is $\boxed{480\text{ nm}}$.

24.25 The thickness of the wedge corresponding to dark bands is $t = \frac{\lambda}{2n}$.

The linear separation between dark bands is x, where

$\frac{t}{x} = \tan\theta \approx \theta = \frac{0.50\text{ mm}}{90\text{ mm}}$.

$x = t\left(\frac{90}{0.50}\right) = \frac{\lambda}{2}\left(\frac{90}{0.50}\right) = 5.9 \times 10^{-5}\text{ m} = \boxed{59\ \mu\text{m}}$.

24.27 The basic relationship is $m\lambda = b\sin\theta \approx b\frac{y}{L}$. Here we need to find the slit width b corresponding to the width of the central maximum, 2y.

$b = 2\,L\frac{\lambda}{2y} = 2\,(1.00\text{ m})\frac{633 \times 10^{-9}\text{ m}}{2.53 \times 10^{-2}\text{ m}} = 5.0 \times 10^{-5}\text{ m} = \boxed{50\ \mu\text{m}}$.

24.29 $\lambda = b\sin\theta \approx b\frac{y}{L}$.

Width of maxima is $2y = \frac{2L\lambda}{b}$, where λ is the wavelength in water $= \lambda_o/n$.

$$\text{Width} = \frac{(2)(4.5\text{ m})(633 \times 10^{-9}\text{ m})}{(0.10 \times 10^{-3}\text{ m})(1.333)} = \boxed{4.3\text{ cm}}.$$

24.31 $\theta = \sin^{-1}\left(\frac{m\lambda}{d}\right)$ $\qquad \Delta\theta = \theta_r - \theta_b = \sin^{-1}\left(\frac{2\lambda_r}{d}\right) - \sin^{-1}\left(\frac{2\lambda_b}{d}\right)$

$$\Delta\theta = \sin^{-1}\left(\frac{(2)(650 \times 10^{-9}\text{ m})}{2.00 \times 10^{-6}\text{ m}}\right) - \sin^{-1}\left(\frac{(2)(450 \times 10^{-9}\text{ m})}{2.00 \times 10^{-6}\text{ m}}\right)$$

$\Delta\theta = 40.5° - 26.7° = \boxed{14.8°}$

24.33 First single–slit minimum occurs at $\lambda = b \sin\theta$.
Fourth order double–slit maximum at $4\lambda = d \sin\theta$.
If the angles are the same then,

$$\sin\theta = \frac{4\lambda}{d} = \frac{\lambda}{b}.$$

$$d = 4b = 4(0.50 \text{ mm}) = \boxed{2.0 \text{ mm}}.$$

24.35 Use the grating equation, noting that the distance between minima can be used to approximate the angle: $\sin\theta \approx \theta \approx \frac{0.50 \text{ m}}{5.0 \text{ m}} = 0.10$ rad.
From $m\lambda = d \sin\theta$ we get

$$d = \frac{\lambda}{\sin\theta} \approx \frac{v/f}{\theta},$$ where v is the speed of sound.

$$d = \frac{(340 \text{ m/s})/(8000/\text{s})}{0.10} = \boxed{0.43 \text{ m}}.$$

24.37 From Fig. 24.45, and also shown below, the path difference of two adjacent rays is $\Delta = d \sin\phi + d \sin\theta$.

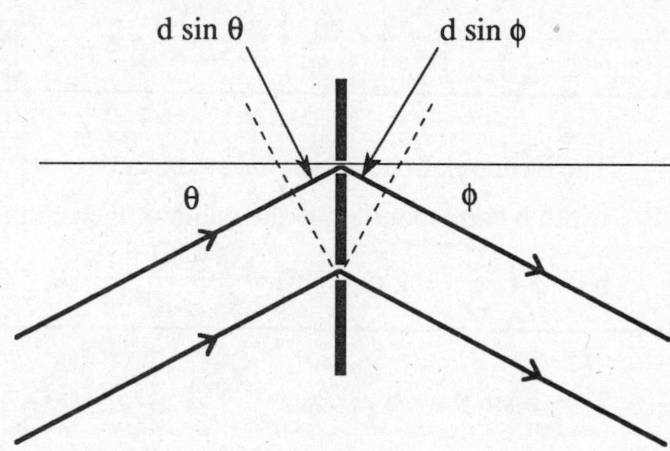

Constructive interference occurs when $\Delta = m\lambda$, where m is an integer. The equation for the maxima in the diffraction pattern becomes
$m\lambda = d\left(\sin\phi + \sin\theta\right)$.

24.39 $$\theta_m = 1.22\frac{\lambda}{D} = 1.22\frac{550 \times 10^{-9} \text{ m}}{5.08 \text{ m}} = \boxed{1.32 \times 10^{-7} \text{ rad}}.$$

24.41 $$\theta = \frac{y}{L} = 1.22\frac{\lambda}{D}.$$

$$L = \frac{yD}{1.22\,\lambda} = \frac{(2.3 \text{ m})(2.0 \times 10^{-3} \text{ m})}{(1.22)(550 \times 10^{-9} \text{ m})} = \boxed{6.9 \text{ km}}.$$

24.43 The minimum angle must be the angle of the dots from the camera lens,

$$\theta = \frac{y}{L} = \theta_m = 1.22\frac{\lambda}{D}. \quad D = \frac{\text{focal length}}{\text{f-number}}.$$

$$\frac{y}{L} = \frac{(1.22)(\lambda)(\text{f-number})}{f},$$

$$\text{f-number} = \frac{fy}{(1.22)\lambda L} = \frac{(135 \times 10^{-3}\text{ m})(0.33 \times 10^{-3}\text{ m})}{(1.22)(550 \times 10^{-9}\text{ m})(100\text{ m})},$$

$$\text{f-number} = 0.66, \text{ which we write as } \boxed{f/0.66}.$$

24.45 The gravitational force provides the centripetal force.

$$m\omega^2 r = \frac{4\pi^2 mr}{T^2} = \frac{GM_E m}{r^2},$$ where r is the distance from the center of the earth to the satellite. T is the length of a day $= 8.64 \times 10^4$ s.

$$r^3 = \frac{GM_E T^2}{4\pi^2},$$

$$r = \sqrt[3]{\frac{GM_E T^2}{4\pi^2}}.$$ The altitude above the earth is $h = r - R_E$, where R_E is the earth's radius $= 6.38 \times 10^6$ m.

$$h = \sqrt[3]{\frac{(6.67 \times 10^{-11}\text{ N·m}^2/\text{kg}^2)(5.98 \times 10^{24}\text{ kg})(8.64 \times 10^4\text{ s})^2}{4\pi^2}} - R_E.$$

$$h = 3.59 \times 10^7\text{ m}.$$

$$\theta_m = 1.22\frac{\lambda}{D} = \frac{\text{object diameter}}{h}.$$

$$\text{Object diameter} = \frac{(1.22)(\lambda)(h)}{D} = \frac{(1.22)(550 \times 10^{-9}\text{ m})(3.59 \times 10^7\text{ m})}{75 \times 10^{-3}\text{ m}},$$

$$\text{Object diameter} = \boxed{320\text{ m}}.$$

24.47 The graph is shown below.

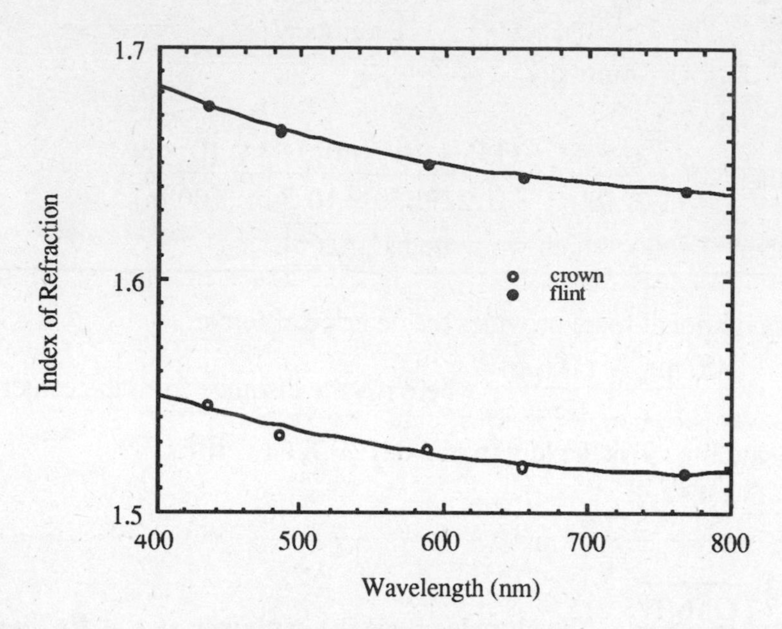

24.49 $I = I_m \cos^2\theta = I_m \cos^2 40° = \boxed{0.59\ I_m}$.

24.51 From the definition of the Brewster angle we have

$\tan\theta_B = \frac{n_t}{n_i}$. $\theta_B = \tan^{-1}\left(\frac{n_t}{n_i}\right) = \tan^{-1}(1.52) = \boxed{56.7°}$.

24.53 (a) From the definition of the Brewster angle we have

$n_{glass} = n_{air} \tan\theta_B = (1.00)\tan(58.9°) = \boxed{1.66}$.

(b) The glass is probably $\boxed{\text{flint glass}}$.

24.55 (a) $I = \boxed{0.50\ I_o}$.

(b) $I = \boxed{0.50\ I_o}$, because there is no further absorption in an ideal polarizer.

(c) $I = (0.50\ I_o)\cos^2 45° = \boxed{0.25\ I_o}$.

24.57 (a) An ideal polarizer would pass 50% of the incident light. This polarizer is only passing $\frac{0.32}{0.50} = 0.64$ as much light as an ideal polarizer. Light passed by the first polarizer is $I_1 = 0.32\ I_o$, where I_o is the intensity of the incident unpolarized light. Light passed by the second polarizer is $I_2 = 0.64\ I_1 \cos^2\theta$, where θ is the angle between the two polarizer, in this case 0°.

$\frac{I_2}{I_o} = (0.64)(0.32)(1) = 0.20 = \boxed{20\%}$.

(b) If $\theta = 32°$, then $\frac{I_2}{I_o} = (0.64)(0.32)\cos^2 32° = (0.64)(0.32)(0.85)^2 = \boxed{15\%}$.

24.59 Use the equation for the maximum in the double slit pattern:

$m\lambda = d \sin\theta$,

$$\lambda = \frac{d}{m}\sin\theta = \frac{0.23 \times 10^{-3}\ \text{m}}{3}\sin 0.51° = 6.8 \times 10^{-7}\ \text{m} = \boxed{680\ \text{nm}}.$$

24.61 Use the equation for the minimum of the double slit pattern.

$$\lambda = \frac{(2y)(b)}{2L},$$ where 2y is the width of the central maximum.

$$\lambda = \frac{(4.7 \times 10^{-2}\ \text{m})(0.050 \times 10^{-3}\ \text{m})}{2(2.0\ \text{m})} = \boxed{590\ \text{nm}}.$$

24.63 $\boxed{\text{Yes}}$ it is possible to do this, because the ice has a smaller index of refraction than does the water. Antireflection layer: $2t = \frac{\lambda}{2}$.

$$t = \frac{\lambda}{4n} = \frac{550\ \text{nm}}{4\,(1.31)} = \boxed{105\ \text{nm}}.$$

24.65 (a) The angle for the longest λ in first order is

$$\theta_1 = \sin^{-1}\left(\frac{\lambda}{d}\right) = \sin^{-1}\left[(700 \times 10^{-9}\ \text{m})(4.0 \times 10^{6}/\text{m})\right] = 16.3°.$$

The angle for the shortest λ in second order is

$$\theta_2 = \sin^{-1}\left(\frac{2\lambda}{d}\right) = \sin^{-1}\left[(2 \times 400 \times 10^{-9}\ \text{m})(4.0 \times 10^{6}/\text{m})\right] = 18.7°.$$

There is $\boxed{\text{no overlap}}$.

(b) The angle for the longest λ in second order is

$$\theta_2 = \sin^{-1}\left(\frac{2\lambda}{d}\right) = \sin^{-1}\left[(2 \times 700 \times 10^{-9}\ \text{m})(4.0 \times 10^{6}/\text{m})\right] = 34.1°.$$

The angle for the shortest λ in third order is

$$\theta_2 = \sin^{-1}\left(\frac{3\lambda}{d}\right) = \sin^{-1}\left[(3 \times 400 \times 10^{-9}\ \text{m})(4.0 \times 10^{6}/\text{m})\right] = 28.7°.$$

$\boxed{\text{Yes}}$, there is overlap.

24.67 (a) From Problem 24.37 we get the equation $m\lambda = d\left(\sin\phi + \sin\theta\right)$.

The two first order beams correspond to $n = \pm 1$.

For $n = +1$,

$$\sin\theta = (\lambda)\left(\frac{1}{d}\right) - \sin\phi = (633 \times 10^{-9}\ \text{m})(2.0 \times 10^{5}/\text{m}) - \sin 15°,$$

$$\theta = \boxed{-7.6°}.$$

For $n = -1$,

$$\sin\theta = -(\lambda)\left(\frac{1}{d}\right) - \sin\phi = -(633\times10^{-9}\ \text{m})(2.0\times10^{5}/\text{m}) - \sin 15°,$$

$\theta = \boxed{-22.7°}$.

Note that for zeroth order (n = 0) $\theta = -\phi$. Thus the angle of the undeviated beam is – 15°. The two first order diffracted beams lie on opposite sides of the undeviated (zeroth order) beam.

(b) Diagrams are shown below.

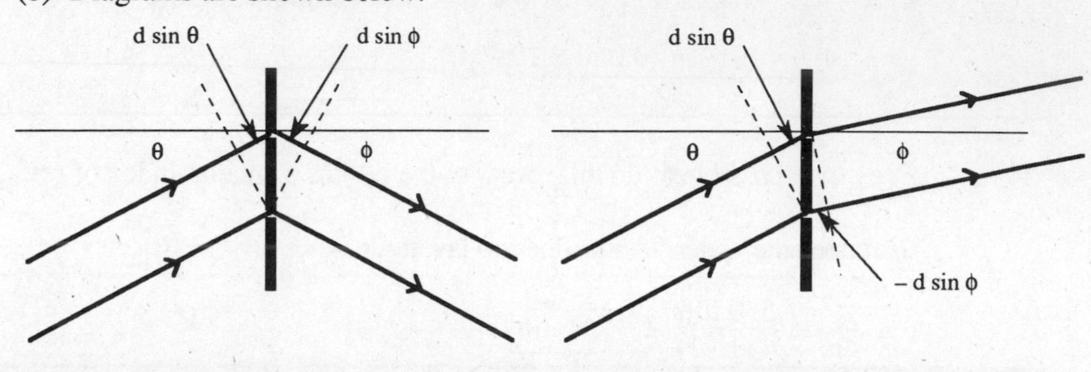

24.69 (a) There will be maximum signal $\boxed{\text{along a north-south line}}$ passing through the center of the array because they will all be in phase.

(b) At 1500 kHz = $1.5\times10^{6}\ \text{s}^{-1}$, the wavelength is

$$\lambda = \frac{c}{f} = \frac{3.00\times10^{8}\ \text{m/s}}{1.5\times10^{6}/\text{s}} = 200\ \text{m}.$$

Since the antennas are spaced 50 m apart along the line, the first antenna will be exactly out of phase with the wave from the third antenna for radiation along the east-west direction. Similarly antenna two is out of phase with antenna four. As a result, there is $\boxed{\text{no signal}}$ radiated along the east-west direction.

24.71 $\theta_m = 1.22\dfrac{\lambda}{D} \approx \dfrac{x}{f}$, where x is the distance between adjacent lines and f is the focal length of the lens. What we are seeking is $\dfrac{1}{x}$, the number of lines per unit length.

$$\frac{1}{x} = \frac{D}{1.22\ \lambda f} = \frac{f/\text{f-number}}{1.22\ \lambda f} = \frac{1}{1.22\ \lambda\ \text{f-number}},$$

$$\frac{1}{x} = \frac{1}{(1.22)(550\times10^{-9}\ \text{m})(1.8)(10^{3}\ \text{mm/m})} = \boxed{830\ \text{lines/mm}}.$$

Chapter 25

RELATIVITY

25.1 $V_A = V_B + 4$ km/h.

$V_{stewardess} = V_s = V_A - 2$ km/h; $V_{passenger} = V_p = V_A + 2$ km/h.

$V_s' = V_s - V_B = \boxed{2 \text{ km/h}}$. $V_p' = V_p - V_B = \boxed{6 \text{ km/h}}$.

25.3 First find the velocity of the boat relative to the shore.

$|v_{Boat}| = \sqrt{v_b^2 + v_r^2} = 5$ km/h $= 1.39$ m/s. The direction of the boat velocity is $\theta = \tan^{-1}\frac{v_r}{v_b} = 53.1°$ in the downstream direction toward the opposite shore.

The passenger's velocity relative to the shore is $\mathbf{v_b} + \mathbf{v_r} = \boxed{0}$.

25.5 Use the formula for velocity addition:

$u = \dfrac{u' + v}{1 + \dfrac{u'v}{c^2}}$. Consider the earth to be the moving frame with speed

$v = 0.50c$, and the rocket to have velocity $u' = -0.70c$.

$$u = \frac{-0.70c + 0.50c}{1 + \dfrac{(-0.70c)(0.50c)}{c^2}} = \frac{-0.20c}{0.65} = \boxed{-0.31c}.$$

25.7 Use the formula for velocity addition:

$u = \dfrac{u' + v}{1 + \dfrac{u'v}{c^2}}$. Consider the spaceship to be the moving frame with speed

$v = 0.80c$. The probe has the velocity $u' = 0.60c$.

$$u = \frac{0.60c + 0.80c}{1 + \dfrac{(0.60c)(0.80c)}{c^2}} = \frac{1.4c}{1.48} = \boxed{0.95c}.$$

25.9 Use the formula for velocity addition.

$u = \dfrac{u' + v}{1 + \dfrac{u'v}{c^2}}$. Consider the spaceship to be the moving frame with speed

$v = 0.70c$. The probe has speed $u' = 0.50c$.

$$u = \frac{1.2c}{1 + 0.35} = \boxed{0.89c}.$$

25.11 Start with the velocity addition formula. Let u and v be speeds that are large in our everyday experience but still small compared to the speed of light. That is, let $u' \ll c$ and $v \ll c$. Then we get

$u = \dfrac{u' + v}{1 + \dfrac{u'v}{c^2}}$, where $\dfrac{u'v}{c^2} << 1$. Thus the denominator is approximately 1. The result is that $\boxed{u = u' + v}$.

25.13 First find the speed of the missile relative to Mars. $v = 0.70c$ and $u' = 0.30c$.

$$u = \frac{u' + v}{1 + \dfrac{u'v}{c^2}} = \frac{0.30c + 0.70c}{1 + (0.30)(0.70)} = 0.826c.$$

Now from the frame of the spaceship, $u' = 0.826c$ and $v = -0.60c$, because the velocity of Mars is in the opposite direction to the velocity of the missile.

$$u = \frac{u' + v}{1 + \dfrac{u'v}{c^2}} = \frac{0.826c - 0.60c}{1 + (0.826)(-0.60)} = \boxed{0.45c}.$$

25.15 The time required for the signal to reach the astronaut is

$$t = \frac{x}{c} = \frac{3.84 \times 10^8 \text{ m}}{3.00 \times 10^8 \text{ m/s}} = 1.28 \text{ s}.$$

He should set his clock to $\boxed{6{:}00{:}01.28}$.

25.17 According to Eq. (25.4)

$\Delta t = \Delta t_o / \sqrt{1 - v^2/c^2}$, where Δt_o is the proper time interval, that is, the time measured on the space ship. Δt is the time measured on the earth.

$$\Delta t_o = \Delta t \sqrt{1 - v^2/c^2} = (20 \text{ y}) \sqrt{1 - (0.85)^2} = \boxed{10.5 \text{ y}}.$$

25.19 We use the time dilation formula, $\Delta t = \Delta t_o / \sqrt{1 - v^2/c^2}$, where Δt_o is the time in the frame of the cosmonaut (2 y) and Δt is the time on earth (8 y).

$$\sqrt{1 - v^2/c^2} = \frac{\Delta t_o}{\Delta t} = \frac{2}{8} = \frac{1}{4}.$$

$$1 - v^2/c^2 = \left(\frac{1}{4}\right)^2 = \frac{1}{16},$$

$$\frac{v^2}{c^2} = 1 - \frac{1}{16} = \frac{15}{16},$$

$$v = \boxed{0.97c}.$$

25.21 We use the time dilation formula, $\Delta t = \Delta t_o / \sqrt{1 - v^2/c^2}$, where Δt_o is the time in the frame of the pion (2.6×10^{-8} s) and Δt is the time in the lab. $v = 0.99c$.

$$\Delta t = \Delta t_o / \sqrt{1 - v^2/c^2} = (2.6 \times 10^{-8} \text{ s}) / \sqrt{1 - (0.99)^2} = \boxed{1.8 \times 10^{-7} \text{ s}}.$$

25.23 We use the time dilation formula, $\Delta t = \Delta t_o/\sqrt{1 - v^2/c^2}$, where Δt_o is the time in the frame of the train (60.0 s) and Δt is the time on the platform (59.5 s). The speed of the train is v = 157 km/h. We must unravel the equation to find the value of c in this universe. First find the speed of the train in m/s.

157 km/h × 1000 m/km × 1 h/3600 s = 43.6 m/s.

$$\sqrt{1 - v^2/c^2} = \Delta t/\Delta t_o,$$

$$1 - v^2/c^2 = (\Delta t/\Delta t_o)^2,$$

$$v^2/c^2 = 1 - (\Delta t/\Delta t_o)^2,$$

$$c^2 = \frac{v^2}{1 - (\Delta t/\Delta t_o)^2},$$

$$c = \frac{v}{\sqrt{1 - (\Delta t/\Delta t_o)^2}} = \frac{43.6 \text{ m/s}}{\sqrt{1 - (59.5/60.0)^2}} = \boxed{339 \text{ m/s}}.$$

25.25 Use the length contraction formula where the proper length is $L_o = 5000$ m.

$$L - L_o\sqrt{1 - v^2/c^2} = (5000 \text{ m})\sqrt{1 - (0.90)^2} = \boxed{2200 \text{ m}}.$$

25.27 Use the length contraction formula to find

$$L/L_o = 0.80 = \sqrt{1 - v^2/c^2},$$

$$1 - v^2/c^2 = (0.80)^2 = 0.64.$$

$$v = c\sqrt{0.36} = \boxed{0.60c}.$$

25.29 We use the length contraction formula in the limit that v << c.

$$L = L_o\sqrt{1 - v^2/c^2} \approx L_o\left(1 - \frac{1}{2}\frac{v^2}{c^2}\right)$$

$$v = 320 \text{ km/h}\frac{1000 \text{ m}}{3600 \text{ s}} = 89 \text{ m/s}.$$

$$\Delta L = L_o - L = L_o\frac{1}{2}\frac{v^2}{c^2} = (100 \text{ m})\frac{1}{2}\frac{(89 \text{ m/s})^2}{(3 \times 10^8 \text{ m/s})^2}$$

$$\Delta L = \boxed{4.4 \times 10^{-12} \text{ m}}.$$

25.31 (a) Use the length contraction formula and the fact the v/c << 1.

$$L = L_o\sqrt{1 - v^2/c^2} \approx L_o\left[1 - \frac{1}{2}\left(\frac{v}{c}\right)^2\right].$$

$$v = 1700 \text{ km/h}\left(\frac{1000 \text{ m/km}}{3600 \text{ s/h}}\right) = 472 \text{ m/s}.$$

$$L - L_o = -\frac{L_o}{2}\left(\frac{v}{c}\right)^2 = -\frac{62.1 \text{ m}}{2}\left(\frac{472 \text{ m/s}}{3.00 \times 10^8 \text{ m/s}}\right)^2 = \boxed{-7.7 \times 10^{-11} \text{ m}}.$$

(b) $L - L_o = 5(-7.7 \times 10^{-11}\text{ m}) = -\dfrac{62.1\text{ m}}{2}\left(\dfrac{v}{3.00 \times 10^8\text{ m/s}}\right)^2,$

$$v = (3.0 \times 10^8\text{ m/s})\sqrt{\frac{(2)(5)(7.7 \times 10^{-11})}{62.1}} = 1060\text{ m/s} = \boxed{3800\text{ km/h}}.$$

25.33 $E = mc^2 = (1.00 \times 10^{-4}\text{ kg})(3.0 \times 10^8\text{ m/s})^2 = \boxed{9 \times 10^{12}\text{ J}}.$

25.35 $E = mc^2 = \text{power} \times \text{time}.$

$$m = \frac{(\text{power})(\text{time})}{c^2} = \frac{(1.0 \times 10^9\text{ W})(3600\text{ s})}{(3.00 \times 10^8\text{ m/s})^2} = 4.0 \times 10^{-5}\text{ kg} = \boxed{40\text{ mg}}.$$

25.37 From the statement of the problem we see that

$p_{relativistic} = 1.05\, p_{nonrelativistic}$

So we use the relativistic momentum formula,

$$p_{relativistic} = \frac{m_o v}{\sqrt{1 - v^2/c^2}} = 1.05\, m_o v,$$

$\sqrt{1 - v^2/c^2} = 1/1.05 = 0.952,$

$1 - v^2/c^2 = 0.907$, so that $v^2/c^2 = 0.0930$. Taking the square root gives

$v/c = 0.305$. Thus $v = \boxed{0.305c}$.

25.39 Use the relativistic mass formula, with $m = 6m_o$,

$$m = 6m_o = \frac{m_o}{\sqrt{1 - v^2/c^2}},$$

$\sqrt{1 - v^2/c^2} = 1/6$, so $v^2/c^2 = 35/36$ and $v = \boxed{0.986c}$.

25.41 Use the relativistic momentum formula,

$$p = \frac{m_o v}{\sqrt{1 - v^2/c^2}},$$

$p^2\left(1 - v^2/c^2\right) = (m_o v)^2,$

$$p^2 = \left(p^2/c^2 + m_o^2\right)v^2 \quad \text{or} \quad v^2 = \frac{p^2}{\left(p^2/c^2 + m_o^2\right)}.$$

If we divide both sides by c^2 we get

$$v^2/c^2 = \frac{p^2/c^2}{\left(p^2/c^2 + m_o^2\right)} = \frac{1}{1 + (m_o c/p)^2},$$

$$v/c = \sqrt{\frac{1}{1 + (m_o c/p)^2}} = \sqrt{\frac{1}{1 + \left(\dfrac{(6.64 \times 10^{-27}\text{ kg})(3.00 \times 10^8\text{ m/s})}{3.75 \times 10^{-19}\text{ kg·m/s}}\right)^2}},$$

$v/c = 0.185.$

$v = 0.185\,c = (0.185)(3.00 \times 10^8\text{ m/s}) = \boxed{5.55 \times 10^7\text{ m/s}}.$

25.43 $KE = m_oc^2 = m_oc^2\left[\dfrac{1}{\sqrt{1-v^2/c^2}} - 1\right],$

$\dfrac{1}{\sqrt{1-v^2/c^2}} = 2$, so $\sqrt{1-v^2/c^2} = 1/2.$

$v/c = \sqrt{3/4}$, $\quad v = \boxed{0.87c}$.

25.45 $KE = 0.50\, m_oc^2 = m_oc^2\left[\dfrac{1}{\sqrt{1-v^2/c^2}} - 1\right], \qquad 1.5 = \dfrac{1}{\sqrt{1-v^2/c^2}},$

$(1 - v^2/c^2) = \dfrac{1}{(1.5)^2}, \qquad v = c\sqrt{1-\dfrac{1}{(1.5)^2}} = \boxed{0.75c}.$

25.47 $E = 5\, m_oc^2 = mc^2 = m_o/\sqrt{1-v^2/c^2},$

$\sqrt{1-v^2/c^2} = 1/5, \quad 1 - v^2/c^2 = 1/25,$

$v^2/c^2 = 1 - 1/25 = 24/25.$

$v = c\sqrt{24/25} = \boxed{0.98c}.$

25.49 $KE = m_oc^2\left[\dfrac{1}{\sqrt{1-v^2/c^2}} - 1\right].$

$\dfrac{1}{\sqrt{1-v^2/c^2}} = 1 + \dfrac{1}{2}\dfrac{v^2}{c^2} + \ldots \approx 1 + \dfrac{1}{2}\dfrac{v^2}{c^2}$, when $v/c << 1$. Then

$KE \approx m_oc^2\left[1 + \dfrac{1}{2}\dfrac{v^2}{c^2} - 1\right] = m_oc^2\left[\dfrac{1}{2}\dfrac{v^2}{c^2}\right] = \boxed{\dfrac{1}{2}m_ov^2}.$

25.51 $f = f_o\sqrt{\dfrac{1+v/c}{1-v/c}}$. Here, $v/c = 0.30$ and $f = 106$ MHz.

$f_o = f\sqrt{\dfrac{1-v/c}{1+v/c}} = (106\text{ MHz})\sqrt{\dfrac{1-0.30}{1+0.30}} = \boxed{78\text{ MHz}}.$

25.53 (a) Longer wavelength indicates $\boxed{\text{moving away}}$.

(b) $\lambda = \dfrac{c}{f} = \dfrac{c}{f_o}\sqrt{\dfrac{1+v/c}{1-v/c}} = \lambda_o\sqrt{\dfrac{1+v/c}{1-v/c}},$

where $\dfrac{v}{c} = \dfrac{4.0\times 10^7\text{ m/s}}{3.0\times 10^8\text{ m/s}} = 0.133.$

$\lambda = (656\text{ nm})\sqrt{\dfrac{1+0.133}{1-0.133}} = \boxed{750\text{ nm}}.$

25.55 The frequency at the earth's surface is f_o; the frequency f detected on a mountain top is smaller due to the gravitational effect.

$$f = f_o\left[1 - \frac{gh}{c^2}\right],$$

$$\Delta f = -\frac{gh}{c^2} f_o = -\frac{(9.8\ \text{m/s}^2)(2000\ \text{m})}{(3.0 \times 10^8\ \text{m/s})^2} f_o = \boxed{-2.2 \times 10^{-13} f_o}.$$

25.57 The rate of motion of the perihelion is constant. We can give that rate as 573 seconds of arc per century or a θ/period where the period is 88 days. Thus

$$\frac{\theta}{88\ \text{d}} = \frac{573\ \text{seconds}}{(100\ \text{y})(365\ \text{d/y})},$$

$\theta = 1.38$ seconds of arc $= 6.7 \times 10^{-6}$ rad.

Now that we have the angular change and the radius (the distance from mercury to the sun) we can calculate the arc length s.

$$s = R\theta = (46 \times 10^9\ \text{m})(6.7 \times 10^{-6}\ \text{rad}) = \boxed{310\ \text{km}}.$$

25.59 From the definition of velocity we get

$$\Delta t = \frac{d}{v} = \frac{6.12\ \text{m}}{0.75c} = 2.72 \times 10^{-8}\ \text{s}.$$

Δt is the lifetime in the frame of the laboratory. The lifetime in the frame of the particles (the proper lifetime) is

$$\Delta t_o = \Delta t\sqrt{1 - v^2/c^2} = (2.72 \times 10^{-8}\ \text{s})\sqrt{1 - (0.75)^2} = 1.8 \times 10^{-8}\ \text{s},$$

$$\Delta t_o = \boxed{18\ \text{ns}}.$$

25.61 First find the kinematic time dilation.

v of earth's surface is $v_E = R_E\omega$.

ω is the angular velocity of the earth = 2π/24 h.

$v_E = (6.38 \times 10^6\ \text{m})(2\pi/24\ \text{h})(1\ \text{h}/3600\ \text{s}) = 464$ m/s.

Let t_s = t of stationary observer, t_E = t of earth observer, and t_p = t of the plane.

$v_p = v_E + 600$ m/s $= 1064$ m/s.

$$t_s = \frac{t_E}{\sqrt{1 - v_E^2/c^2}},\ \text{so}\ t_E = t_s\sqrt{1 - v_E^2/c^2} \approx t_s\left[1 - \frac{1}{2}\left(\frac{v_E}{c}\right)^2\right].$$

Similarly, $t_p \approx t_s\left[1 - \frac{1}{2}\left(\frac{v_p}{c}\right)^2\right]$.

$\Delta t = t_p - t_E \approx \frac{1}{2}\left(\frac{v_E^2 - v_p^2}{c^2}\right)t_E$. Note the here we have substituted t_E for t_s because the difference is negligible.

Inserting the values for v_E, v_p, and t_E gives $\Delta t = -330$ ns.

Now find the gravitational time dilation effect.

$$t_E = \frac{1}{f_E}\ \text{and}\ t_p = \frac{1}{f_p} = \frac{1}{f_E(1 - gh/c^2)} \approx t_E\left(1 + \frac{gh}{c^2}\right).$$

$$\Delta t = t_p - t_E = \left(\frac{gh}{c^2}\right) t_E = \frac{(9.8 \text{ m/s}^2)(12 \times 10^3 \text{ m})}{(3 \times 10^8 \text{ m/s})^2}(18 \text{ h})(3600 \text{ s/h}) = 85 \text{ ns}.$$

$$\Delta t_{total} = \Delta t_{kinematic} + \Delta t_{gravitational} = -330 \text{ ns} + 85 \text{ ns} = \boxed{-245 \text{ ns}}.$$

25.63 We start with the formula for length contraction:

$l = l_o\sqrt{1-(v/c)^2}$, which can be solved for $v/c = \sqrt{1-(l/l_o)^2}$.

The momentum becomes $p = \dfrac{mv}{\sqrt{1-(v/c)^2}} = \dfrac{(mc)(v/c)}{\sqrt{1-v^2/c^2}}$,

$$p = \frac{(mc)(\sqrt{1-(l/l_o)^2})}{l/l_o}$$

$$p = \frac{(9.11 \times 10^{-31} \text{ kg})(3.00 \times 10^8 \text{ m/s})\sqrt{1-(7.83/10.0)^2}}{7.83/10.0}$$

$$p = \boxed{2.17 \times 10^{-22} \text{ kg}\cdot\text{m/s}}.$$

Chapter 26

THE DISCOVERY OF ATOMIC STRUCTURE

26.1 Arrange in a square.

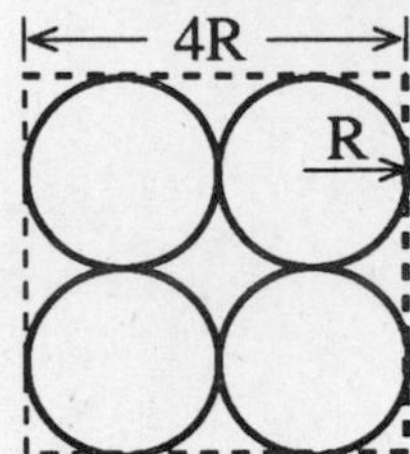

Four pennies occupy area of square of edge length 4R, where R is the radius of a penny. Area per penny is $A = (4R)^2/4 = 4R^2$.

Arrange in a hexagon.

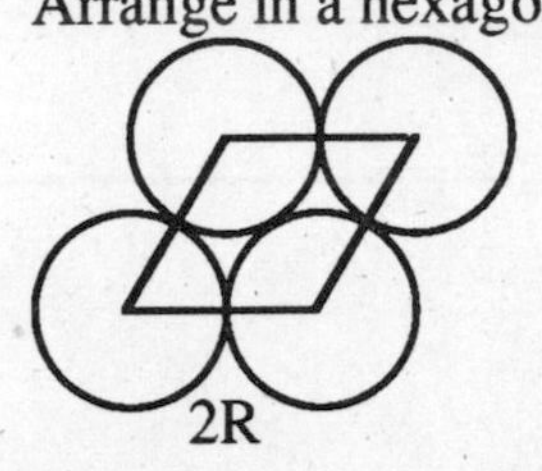

Notice that the pennies can be grouped into parallelograms of four pennies. Each parallelogram drawn to the centers of the pennies contains one penny. The edge lengths of the parallelogram are 2R and the acute angle is 60°. Thus the area taken up by each penny is $A = (2R)^2 \sin 60° = (0.87)4R^2$ and is less than the space needed for the square array.

26.3 (a) The ratio $\frac{1.37}{2.15} = 0.637$. It is not a ratio of small whole numbers. Results are probably $\boxed{\text{untrue}}$.

(b) The ratio $\frac{1.74}{2.61} = 0.6667 = \frac{2}{3}$. The results are in the ratio of two small whole numbers. The results are probably $\boxed{\text{true}}$.

26.5 If M is the atomic mass, then the ratio of the mass m plated in time t to the atomic mass is $m/M = It/F$,

where I is the current and F is the Faraday.

$$m = \frac{ItM}{F} = \frac{(0.500\text{ A})(2400\text{ s})(108\text{ g/mol})}{96485\text{ C/mol}} = \boxed{1.34\text{ g}}.$$

26.7 If M is the atomic mass, then the ratio of the mass m plated in time t to 1/x of the atomic mass is

$m/(M/x) = It/F$,

where I is the current and F is the Faraday. We use M/x because the ions are assumed to have a charge of xe.

$$m = \frac{It(M/x)}{F}.$$

The number of charges per ion is x:

$$x = \frac{ItM}{mF} = \frac{(0.25\text{ A})(2400\text{ s})(63.6\text{ g/mol})}{(0.396\text{ g})(96485\text{ C/mol})} = 1.0. \quad \boxed{\text{Ions singly charged.}}$$

26.9 First find the number of moles released. Remember O is doubly charged.

$$\frac{m}{M} = \frac{It}{xF} = \frac{(6.0\text{ A})(3600\text{ s})}{(2)(96485\text{ C/mol})} = 0.112\text{ mol of O atoms.}$$

Because the gaseous oxygen is diatomic, there is only 0.056 mol of O_2.
At STP one mol of a gas occupies 22.4 L (see Problem 25.9), so the volume of gas liberated is

$$V = 0.056\text{ mol} \times 22.7\text{ L/mol} = \boxed{1.3\text{ L}}.$$

26.11 Graph shown below.

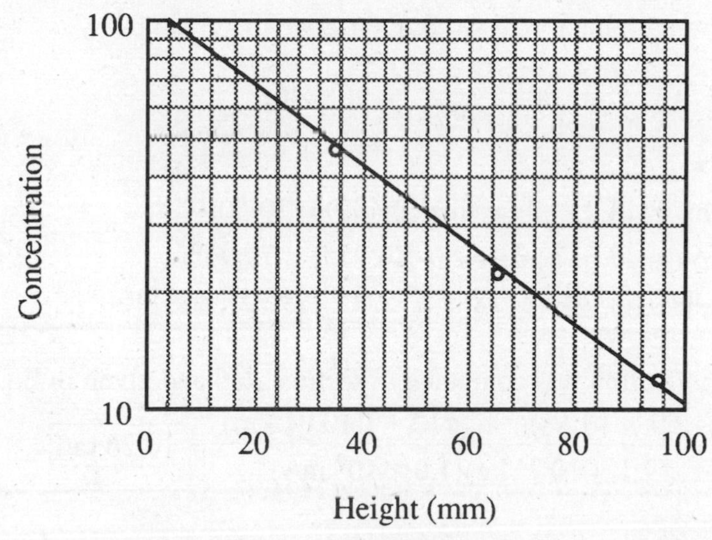

The points do fit a reasonably straight line. A straight line on semilograrithmic graph means that the function is exponential.

26.13 Start with $n = n_o\, e^{-\lambda h}$. Then let $h = 0.693/\lambda$.

$$n = n_o\, e^{-\lambda(0.693/\lambda)} = n_o e^{-0.693} = n_o(0.50) = \boxed{n_o/2}.$$

26.15 Given $n(h) = n(0)\, e^{-3g(\rho - \rho')Vh/2\,\overline{KE}}$ and $\overline{KE} = 3RT/2N_A$.

$$\ln\left(\frac{n(h)}{n(0)}\right) = \frac{-3g(\rho - \rho')Vh}{2\,\overline{KE}} = \frac{3g(\rho' - \rho)Vh\, 2N_A}{2(3RT)},$$

$$N_A = \frac{RT}{g(\rho' - \rho)Vh} \ln\left(\frac{n(h)}{n(0)}\right).$$

26.17 The number of atoms in a cubic centimeter is the density in g/cm^3 divided by the atomic mass in g/mol and multiplied by Avogadro's number:

$$N = \frac{\rho N_A}{M} = \frac{(2.23\text{ g/cm}^3)(6.02 \times 10^{23}\text{/mol})}{28.09\text{ g/mol}} = \boxed{4.78 \times 10^{22}\text{/cm}^3}.$$

26.19 Use the Bragg equation $n\lambda = 2d \sin \theta$,

$$d = \frac{n\lambda}{2 \sin \theta} = \frac{0.709 \times 10^{-10} \text{ m}}{2 \sin (23.2°/2)} = \boxed{1.76 \times 10^{-10} \text{ m}}.$$

26.21 Use the Bragg equation $n\lambda = 2d \sin \theta$,

$$d = \frac{n\lambda}{2 \sin \theta} = \frac{2 \times 0.709 \times 10^{-10} \text{ m}}{2 \sin (28.50°/2)}, \qquad d = \boxed{2.88 \times 10^{-10} \text{ m}}.$$

26.23 Use the Bragg equation $n\lambda = 2d \sin \theta$, with $n = 1$.

$$\lambda = 2d \sin \theta = 2(4.07 \times 10^{-10} \text{ m}) \sin (24.0°/2) = \boxed{1.69 \times 10^{-10} \text{ m}}.$$

26.25 Use the Bragg equation $n\lambda = 2d \sin \theta$. Rearrange to get $\sin \theta$.

$$\sin \theta = \frac{n\lambda}{2d} = n\frac{0.709 \times 10^{-10} \text{ m}}{2 \times 1.44 \times 10^{-10} \text{ m}} = n(0.246).$$

$\sin \theta < 1$ for $n = 1, 2, 3$, and 4.

$\theta_1 = \sin^{-1}(0.246) = 14.25$; $\theta_2 = \sin^{-1}(0.492) = 29.50$;

$\theta_3 = \sin^{-1}(0.739) = 47.61°$; $\theta_4 = \sin^{-1}(0.985) = 79.97°$.

The allowed values for 2θ are $\boxed{28.5°}$, $\boxed{59.0°}$, $\boxed{95.2}$, and $\boxed{160°}$.

26.27 The magnetic deflection was computed in section 26.6 and given in Eq. (26.6):

$$\theta_B = \frac{eBl}{mv} = \frac{(1.60 \times 10^{-19})(1.0 \times 10^{-3} \text{ T})(0.015 \text{ m})}{(9.1 \times 10^{-31} \text{ kg})(1.0 \times 10^{7} \text{ m/s})} = \boxed{0.26 \text{ rad}}.$$

26.29 The kinetic energy of the electrons is $\frac{1}{2} mv^2 = eV_o$,

The deflection is $\theta_E = \dfrac{eEl}{mv^2} = \dfrac{eVl}{d\, 2\, eV_o}$.

The deflection is $\theta_E = \tan \theta_E = 3.00 \text{ cm}/30.0 \text{ cm} = 0.100$ rad.

$$V = \frac{\theta_E\, d\, 2\, V_o}{l} = \frac{(0.100)(4.00 \times 10^{-3} \text{ m})(2 \times 19 \times 10^{3} \text{ V})}{2.00 \times 10^{-2} \text{ m}} = \boxed{760 \text{ V}}.$$

26.31 If the electric and magnetic deflections are the same, then the speed of the particle is $v = E/B$. We can then use Eq. (26.7) to evaluate E.

$$V = Ed = \frac{eB^2ld}{m\theta_E},$$

$$V = \frac{(1.6 \times 10^{-19} \text{ C})(1.00 \times 10^{-2} \text{ T})^2(0.0200 \text{ m})(4.00 \times 10^{-3} \text{ m})}{(9.11 \times 10^{-31} \text{ kg})(0.26 \text{ rad})},$$

$$V = \boxed{5.4 \text{ kV}}.$$

26.33 Centripetal force provided by the magnetic force: $\frac{mv^2}{r} = evB$.

$$r = \frac{mv}{eB} = \frac{(9.1 \times 10^{-31}\ \text{kg})(2.0 \times 10^{7}\ \text{m/s})}{(1.6 \times 10^{-19}\ \text{C})(0.50\ \text{T})} = 2.3 \times 10^{-4}\ \text{m} = \boxed{0.23\ \text{mm}}.$$

26.35 Centripetal force provided by the magnetic force: $\frac{mv^2}{r} = evB$.

$$v = \frac{erB}{m} = \frac{(2 \times 1.60 \times 10^{-19}\ \text{C})(4.26\ \text{m})(6.40 \times 10^{-2}\ \text{T})}{(7294)(9.11 \times 10^{-31}\ \text{kg})},$$

$$v = \boxed{1.31 \times 10^{7}\ \text{m/s}}.$$

26.37 Calculate the speed from $KE = \frac{1}{2}mv^2$, $v = \sqrt{\frac{2\,KE}{m}} = 5.13 \times 10^{6}$ m/s.

Centripetal force is provided by the magnetic force: $\frac{mv^2}{r} = qvB$, so

$$r = \frac{mv}{qB} = \frac{6.6 \times 10^{-27}\ \text{kg} \times 5.13 \times 10^{6}\ \text{m/s}}{3.2 \times 10^{-19}\ \text{C} \times 1.00\ \text{T}} = \boxed{0.11\ \text{m}}.$$

26.39 After thirty min = 10 half-life, the activity will be $\left(\frac{1}{2}\right)^{10}$ of the initial activity.

$$A = \frac{1.0 \times 10^{7}\ \text{Bq}}{2^{10}} = \frac{1.0 \times 10^{7}\ \text{Bq}}{1024} = \boxed{9.8 \times 10^{3}\ \text{Bq}}.$$

26.41 (a) From the graph $\boxed{t_{1/2} = 3.5\ \text{h}}$.

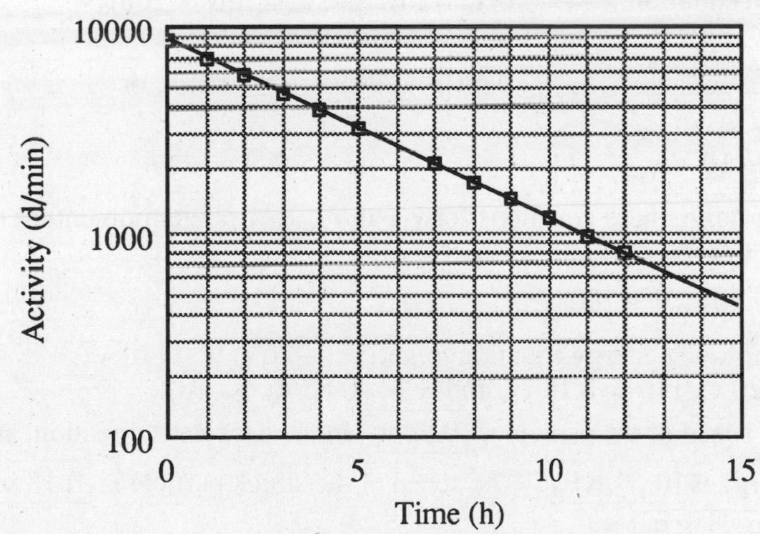

(b) The disintegration constant is $\lambda = \frac{0.693}{t_{1/2}} = \boxed{0.20/\text{h}}$.

(c) At t = 6 h, A = $\boxed{\text{2600 disintegrations/min}}$.

26.43 Half-life = T = 5700 y.

$N = N_o e^{-0.693t/T} = N_o/3$. Divide by N_o and take logarithms of both sides.

$-0.693\, t/T = \ln(1/3) = -1.10.$ $t = \boxed{9000\text{ y}}$.

26.45 Refer to Example 26.8 where we found that the distance of closest approach is

$$b = \frac{4kZe^2}{m_\alpha v^2} = \frac{(4 \times 9 \times 10^9\ \text{N}\cdot\text{m}^2/\text{C}^2)(82)(1.6 \times 10^{-19}\ \text{C})^2}{(6.6 \times 10^{-27}\ \text{kg})(1.8 \times 10^7\ \text{m/s})^2},$$

$b = \boxed{3.5 \times 10^{-14}\text{ m}}$.

26.47 Refer to Example 26.8 where we found that the distance of closest approach is

$$b = \frac{4kZe^2}{m_\alpha v^2}.$$

Gold: $$b = \frac{4kZe^2}{m_\alpha v^2} = \frac{(4 \times 9 \times 10^9\ \text{N}\cdot\text{m}^2/\text{C}^2)(79)(1.6 \times 10^{-19}\ \text{C})^2}{(6.6 \times 10^{-27}\ \text{kg})(1.70 \times 10^7\ \text{m/s})^2},$$

$b_{gold} = \boxed{3.8 \times 10^{-14}\text{ m}}$.

Copper: $$b = \frac{4kZe^2}{m_\alpha v^2} = \frac{(4 \times 9 \times 10^9\ \text{N}\cdot\text{m}^2/\text{C}^2)(29)(1.6 \times 10^{-19}\ \text{C})^2}{(6.6 \times 10^{-27}\ \text{kg})(1.70 \times 10^7\ \text{m/s})^2},$$

$b_{copper} = \boxed{1.4 \times 10^{-14}\text{ m}}$.

26.49 The count rate is proportional to $1/r^2$, where r is the distance between the source and the detector. If the detector is moved back from 5 cm to 10 cm from the source, the count rate will decrease by a factor of 4. The count rate is

$\frac{1}{4}\ 4320/\text{min} = \boxed{1080/\text{min}}$.

26.51 Momentum equation: $mv_o = mv + MV = mv + 10m_oV.$

So, $v_o = v + 10V.$

Energy equation: $\frac{1}{2}mv_o^2 = \frac{1}{2}mv^2 + \frac{1}{2}10mV^2$, or

$v_o^2 = v^2 + 10\,V^2.$

We need to solve these equations for v and V. Rearrange momentum equation and square to get

$v^2 = v_o^2 - 20\,v_oV + 100\,V^2.$

Insert this into the energy equation to get $0 = -20\,v_oV + 110\,V^2$.

Then $V = 20\,v_o/110 = 0.18\,v_o$, and $v = -0.82\,v_o$.

The recoil speed of the particle is $\boxed{0.82\,v_o}$ in the backward direction, and its kinetic energy is $\boxed{0.67\,KE_o}$. The speed of the struck particle is $\boxed{0.18\,v_o}$ and its energy is $\boxed{0.33\,KE_o}$.

Chapter 27

ORIGINS OF THE QUANTUM THEORY

27.1 (a) Use the Balmer formula to get

$$\lambda = 364.56 \text{ nm} \frac{n^2}{n^2 - 2^2} = 364.56 \text{ nm} \frac{16}{16-4} = \boxed{486.08 \text{ nm}}.$$

This line is in the $\boxed{\text{blue}}$.

(b) $\lambda = 364.56 \text{ nm} \frac{25}{25-4} = \boxed{434.00 \text{ nm}}$. This line is $\boxed{\text{violet}}$.

27.3 The wavelengths of the first four lines in the Balmer series correspond to n = 3, 4, 5, 6 in the Balmer formula.

$$\lambda_{(3)} = 364.56 \text{ nm} \frac{n^2}{n^2 - 2^2} = 364.56 \text{ nm} \frac{9}{9-4} = \boxed{656.21 \text{ nm}},$$

$$\lambda_{(4)} = 364.56 \text{ nm} \frac{16}{16-4} = \boxed{486.08 \text{ nm}},$$

$$\lambda_{(5)} = 364.56 \text{ nm} \frac{25}{25-4} = \boxed{434.00 \text{ nm}},$$

$$\lambda_{(6)} = 364.56 \text{ nm} \frac{36}{36-4} = \boxed{410.13 \text{ nm}}.$$

27.5 From the Wien displacement law we have $\lambda_m T = 2.90 \times 10^{-3}$ m·K.
Here $\lambda_m = 475$ nm.

$$T = \frac{2.90 \times 10^{-3} \text{ m·K}}{475 \times 10^{-9} \text{ m}} = \boxed{6110 \text{ K}}.$$

27.7 From the Wien displacement law we have $\lambda_m T = 2.90 \times 10^{-3}$ m·K.
Here T = 5100 K, so

$$\lambda_m = \frac{2.90 \times 10^{-3} \text{ m·K}}{5100 \text{ K}} = \boxed{569 \text{ nm}}.$$

27.9 From the Wien displacement law we have $\lambda_m T = 2.90 \times 10^{-3}$ m·K.
Here T = 3200 K, so

$$\lambda_m = \frac{2.90 \times 10^{-3} \text{ m·K}}{3200 \text{ K}} = \boxed{910 \text{ nm}}.$$

27.11 $E = hf = \frac{hc}{\lambda} = \frac{(6.63 \times 10^{-34} \text{ J·s})(3.00 \times 10^{8} \text{ m/s})}{955 \times 10^{-9} \text{ m}} = \boxed{2.08 \times 10^{-19} \text{ J}}$.

27.13 $\lambda = \frac{hc}{E} = \frac{(6.63 \times 10^{-34} \text{ J·s})(3.00 \times 10^{8} \text{ m/s})}{4.82 \times 10^{-19} \text{ J}} = \boxed{413 \text{ nm}}$.

27.15 (a) Because the photon has no mass, the relativistic expression for energy, when combined with Planck's hypothesis gives: $p = \frac{E}{c} = \boxed{\frac{hf}{c}}$.

(b) F = weight = mg, also $F = \frac{\Delta p}{\Delta t} = \frac{n(2p)}{\Delta t}$.

Upon reflection the momentum goes from p to – p, so $\Delta p = 2p$. n represents the number of photons. From Newton's second law we get

$$n(\text{per second}) = \frac{mg}{2p} = \frac{mgc}{2hf} = \frac{mg\lambda}{2h}.$$

$$n = \frac{(0.50 \times 10^{-3}\text{ kg})(9.81\text{ m/s}^2)(475 \times 10^{-9}\text{ m})}{2(6.63 \times 10^{-34}\text{ J·s})} = \boxed{1.76 \times 10^{24}/\text{s}}.$$

(c) $P = nE = n\frac{hc}{\lambda} = 1.76 \times 10^{24}/\text{s}\ \frac{(6.63 \times 10^{-34}\text{ J·s})(3.00 \times 10^{8}\text{ m/s})}{475 \times 10^{-9}\text{ m}}$,

$P = \boxed{737\text{ kW}}$.

(d) This experiment would require an extremely intense beam of light with cross–sectional area of 3 cm × 3 cm and is not a practical one.

27.17 $\lambda = \frac{hc}{E} = \frac{(4.136 \times 10^{-15}\text{ eV·s})(3.00 \times 10^{8}\text{ m/s})}{2.00\text{ eV}} = \boxed{620\text{ nm}}$.

27.19 The shortest wavelength corresponds to the largest energy.

$$\lambda_{min} = \frac{hc}{E_{max}} = \frac{(4.136 \times 10^{-15}\text{ eV·s})(3.00 \times 10^{8}\text{ m/s})}{35.7 \times 10^{3}\text{ eV}} = \boxed{3.48 \times 10^{-11}\text{ m}}.$$

27.21 $E = \frac{hc}{\lambda} = \frac{(4.136 \times 10^{-15}\text{ eV·s})(2.998 \times 10^{8}\text{ m/s})}{589.3 \times 10^{-9}\text{ m}} = \boxed{2.104\text{ eV}}$.

27.23 $\lambda_{th} = \frac{hc}{\phi} = \frac{(4.136 \times 10^{-15}\text{ eV·s})(3.00 \times 10^{8}\text{ m/s})}{4.74\text{ eV}} = \boxed{262\text{ nm}}$.

27.25 (a) $\lambda = \frac{hc}{E} = \frac{(4.136 \times 10^{-15}\text{ eV·s})(3.00 \times 10^{8}\text{ m/s})}{90.0 \times 10^{3}\text{ eV}}$,

$\lambda = \boxed{1.38 \times 10^{-11}\text{ m}}$.

(b) $f = \frac{c}{\lambda} = \frac{E}{h} = \frac{90.0 \times 10^{3}\text{ eV}}{4.136 \times 10^{-15}\text{ eV·s}} = \boxed{2.18 \times 10^{19}\text{ Hz}}$.

27.27 (a) From the photoelectric equation we get

$$\phi = \frac{hc}{\lambda} - KE_{max} = \frac{(4.136 \times 10^{-15}\text{ eV·s})(3.00 \times 10^{8}\text{ m/s})}{546.1 \times 10^{-9}\text{ m}} - 0.13\text{ eV},$$

$\phi = 2.27 \text{ eV} - 0.13 \text{ eV} = \boxed{2.14 \text{ eV}}$.

(b) The material is probably $\boxed{\text{cesium}}$.

27.29 (a) $E = 10\dfrac{hc}{\lambda} = 10\dfrac{6.63 \times 10^{-34} \text{ J·s} \times 3.00 \times 10^{8} \text{ m/s}}{600 \times 10^{-9} \text{ m}}$,

$E = \boxed{3.31 \times 10^{-18} \text{ J}}$,

(b) Convert the energy into change in gravitational potential, mgh. Here h is the height not the Planck constant.

$$h = \frac{1000\ E}{mg} = \frac{1000 \times 3.31 \times 10^{-18} \text{ J}}{2.0 \times 10^{-9} \text{ kg} \times 9.8 \text{ m/s}^2} = \boxed{169 \text{ nm}}.$$

27.31 Use the equation for the orbital radius found as Eq. (27.10), $r_n = \dfrac{\varepsilon_o n^2 h^2}{\pi m e^2}$.

Here use n = 2. The result is $r_2 = \dfrac{\varepsilon_o\, 4\, h^2}{\pi m e^2} = \boxed{2.12 \times 10^{-10} \text{ m}}$.

27.33 The angular momentum is given by $L = mvr$. The tangential speed is obtained as

$v = \dfrac{2\pi r}{T} = 2\pi r f$. So, $\quad L = m\, 2\pi r^2 f = n\dfrac{h}{2\pi}$.

$$n = \frac{m\,(2\pi)^2\, r^2 f}{h} = \frac{5.00 \times 10^{-3} \text{ kg}\, (2\pi)^2\, (0.10 \text{ m})^2 \times 3.0\text{/s}}{6.63 \times 10^{-34} \text{ J·s}} = \boxed{8.9 \times 10^{30}}.$$

27.35 We can compute the wavelength from $\lambda = \dfrac{hc}{E}$, where E is the energy difference of $E = E_2 - E_1$. Here $E_2 = 0$ and E_1 is the ground state $= -13.6$ eV. Then the value for E becomes 13.6 eV.

$$\lambda = \frac{(4.136 \times 10^{-15} \text{ eV·s})(3.00 \times 10^{8} \text{ m/s})}{13.6 \text{ eV}} = \boxed{91.2 \text{ nm}}.$$

27.37 Use the Rydberg formula: $\dfrac{1}{\lambda} = R\left(\dfrac{1}{n_1^2} - \dfrac{1}{n_2^2}\right)$.

Let $n_1 = 4$ and $n_2 = \infty$. Then $\lambda = 16/R = \boxed{1460 \text{ nm}}$.

27.39 The Rydberg constant was defined from the Bohr equations as $R = \dfrac{me^4}{8\varepsilon_o^2 h^3 c}$,

$$R = \frac{9.1095 \times 10^{-31} \text{ kg}\, (1.602 \times 10^{-19} \text{ C})^4}{8(8.854 \times 10^{-12} \text{ F/m})^2 (6.626 \times 10^{-34} \text{ J·s})^3\, 2.998 \times 10^{8} \text{ m/s}},$$

$R = \boxed{1.097 \times 10^{7}\text{/m}}$.

27.41 (a) At room temperature we can consider that all of the atoms have their electrons in the lowest state. The absorption then corresponds to absorption from the n = 1 level to the n = 2 level, the next higher level.

$$E = \frac{hc}{\lambda} = E_2 - E_1 = -13.6\text{ eV}(1/4 - 1) = \boxed{10.2\text{ eV}}.$$

(b) $$\lambda = \frac{hc}{E} = \frac{(4.14 \times 10^{-15}\text{ eV}\cdot\text{s})(3.00 \times 10^8\text{ m/s})}{10.2\text{ eV}} = \boxed{122\text{ nm}}.$$

27.43 (a) The angular momentum is given by

$L = mvr = nh/2\pi$. For the second Bohr orbit n = 2 and $r = 4r_1$.

$$v = \frac{h}{m4r_1\pi} = \frac{6.63 \times 10^{-34}\text{ J}\cdot\text{s}}{(9.11 \times 10^{-31}\text{ kg})(4 \times 5.29 \times 10^{-11}\text{ m})(\pi)},$$

$$v = \boxed{1.09 \times 10^6\text{ m/s}}.$$

(b) $v/c = \boxed{3.65 \times 10^{-3}}$. Non relativistic treatment is reasonable.

27.45 (a) Moseley's law may be expressed as $E = 10.2\,(Z-1)^2$ eV. Here Z = 64 so $E = 10.2\,(64-1)^2\text{ eV} = \boxed{40.5\text{ keV}}$.

(b) For Z = 90 we get $E = 10.2\,(90-1)^2\text{ eV} = \boxed{80.8\text{ keV}}$.

27.47 (a) From Moseley's law we get

$$\lambda = \frac{4}{3R(Z-1)^2} = \frac{4}{3(1.097 \times 10^7\text{ m})(74-1)^2} = \boxed{2.28 \times 10^{-11}\text{ m}}.$$

(b) $$\lambda = \frac{4}{3(1.097 \times 10^7/\text{m})(42-1)^2} = \boxed{7.23 \times 10^{-11}\text{ m}}.$$

27.49 The mimimum accelerating voltage must equal the x ray energy divided by e.

$$V = \frac{\text{energy}}{e} = \frac{hc}{\lambda e} = \frac{4.14 \times 10^{-15}\text{ eV}\cdot\text{s} \times 3.00 \times 10^8\text{ m/s}}{0.048 \times 10^{-9}\text{ m} \times e} = \boxed{25.9\text{ kV}}.$$

27.51 Moseley's law may be expressed as $E = 10.2\,(Z-1)^2\text{ eV} = 8.07$ keV,

$$Z - 1 = \sqrt{\frac{8070}{10.2}} = 28.$$ Z = 29, the element is $\boxed{\text{copper}}$.

27.53 Moseley's law may be expressed as $\lambda = \frac{4}{3R(Z-1)^2}$. If $\lambda = 0.079$ nm, then

$$Z - 1 = \sqrt{\frac{4}{3(1.097 \times 10^7/\text{m})(0.079 \times 10^{-9}\text{ m})}},$$

Z = 1 + 39 = 40. The element is $\boxed{\text{zirconium}}$.

27.55 From the Rydberg formula we get

$$\frac{1}{\lambda} = R\left(\frac{1}{n_1^2} - \frac{1}{n_2^2}\right) = R\left(\frac{1}{2^2} - \frac{1}{3^2}\right) = R\,(0.139) = 1.52 \times 10^6/\text{m}.$$

$\lambda = 656$ nm.

For n = 2, the Bohr radius is $r_2 = 4r_1 = 4\,(0.0529 \text{ nm}) = 0.212$ nm.

$$\frac{\lambda}{r_2} = \frac{656 \text{ nm}}{0.212 \text{ nm}} = \boxed{3.09 \times 10^3}.$$

27.57 (a) From the Rydberg formula we get

$$E = 13.6 \text{ eV}\left(\frac{1}{n_1^2} - \frac{1}{n_2^2}\right).$$ Here we let $n_1 = 3$ and $n_2 = \infty$.

$$E = 13.6 \text{ eV}/9 = \boxed{1.51 \text{ eV}}.$$

(b) $E = KE = \boxed{0}$.

(c) $f = \Delta E/h = 1.51 \text{ eV}/4.14 \times 10^{-15} \text{ eV}\cdot\text{s} = \boxed{3.65 \times 10^{14} \text{ Hz}}$.

27.59 Start with $L = mvr = n\dfrac{h}{2\pi}$.

The centripetal force is supplied by the electrostatic force:

$$\frac{mv^2}{r} = \frac{e^2}{4\pi\varepsilon_o r^2}.$$ Upon substituting for mvr we get

$$\frac{mv^2}{r} = \frac{(mvr)^2}{mr^3} = \frac{n_2h_2}{4\pi^2 mr^3} = \frac{e^2}{4\pi\varepsilon_o r^2}.$$ This reduces to $r_n = \dfrac{\varepsilon_o n^2h^2}{\pi me^2}$.

The total energy is $E = KE + PE = \frac{1}{2}mv^2 - \dfrac{e^2}{4\pi\varepsilon_o r}$.

$$E = \frac{1}{2}\left(\frac{e^2}{4\pi\varepsilon_o r_n}\right) - \frac{e^2}{4\pi\varepsilon_o r_n} = -\frac{e^2}{8\pi\varepsilon_o r_n}.$$

Substituting for r_n we get $E_n = \dfrac{-me^4}{8\varepsilon_o^2 h^2 n^2}$.

The energy of the light emitted in a transition from an upper state E_2 to a lower state E_1 is $\Delta E = hf = \dfrac{hc}{\lambda} = E_2 - E_1$. So,

$$\frac{1}{\lambda} = \frac{1}{hc}(E_2 - E_1),$$

$$\frac{1}{\lambda} = \frac{me^4}{8\varepsilon_o^2 h^3 c}\left(\frac{1}{n_1^2} - \frac{1}{n_2^2}\right).$$

Chapter 28

QUANTUM MECHANICS

28.1 From the equation for Compton scattering we get

$$\lambda' = \lambda + \left(\frac{h}{mc}\right)(1 - \cos\theta) = 0.025 \text{ nm} + 2.426 \times 10^{-12} \text{ m} (1 - \cos 60°),$$

$$\lambda' = 2.62 \times 10^{-11} \text{ m}.$$

$$f' = \frac{c}{\lambda'} = \frac{2.998 \times 10^{8} \text{ m/s}}{2.62 \times 10^{-11} \text{ m}} = \boxed{1.14 \times 10^{19} \text{ Hz}}.$$

28.3 From the equation for Compton scattering we get

$\lambda' = \lambda + \left(\frac{h}{mc}\right)(1 - \cos\theta)$. The wavelength λ can be obtained from the rule given in Chapter 27 that $\lambda = \frac{hc}{E} = \frac{1240 \text{ eV}\cdot\text{nm}}{E}$. So,

$$\lambda' = \lambda + \left(\frac{h}{mc}\right) = \frac{1240 \text{ eV}\cdot\text{nm}}{10^{5} \text{ eV}} + 2.426 \times 10^{-12} \text{ m} = 1.48 \times 10^{-11} \text{ m}.$$

$$E' = \frac{hc}{\lambda} = \frac{1240 \text{ eV}\cdot\text{nm}}{0.0148 \text{ nm}} = \boxed{83.6 \text{ keV}}.$$

28.5 (a) From the equation for Compton scattering we get

$$\lambda' = \lambda + \left(\frac{h}{mc}\right)(1 - \cos\theta),$$

$$\lambda' = \frac{1240 \text{ eV}\cdot\text{nm}}{17.2 \times 10^{3} \text{ eV}} + 2.426 \times 10^{-12} \text{ m} (1 - \cos 120°) = 7.57 \times 10^{-11} \text{ m}.$$

$$E' = \frac{1240 \text{ eV}\cdot\text{nm}}{0.0757 \text{ nm}} = \boxed{16.4 \text{ keV}}.$$

(b) Electron recoil energy is $E - E' = \boxed{0.8 \text{ keV}}$.

28.7 The energy of the incident photon is $E = \frac{hc}{\lambda} = \frac{1240 \text{ eV}\cdot\text{nm}}{0.0023 \text{ nm}} = 5.39 \times 10^{5}$ eV.

The energy of the scattered photon is

$E' = (1 - 0.024)\, E = 5.26 \times 10^{5}$ eV, and its wavelength is

$\lambda' = \frac{hc}{E'} = 0.00236$ nm. The change in wavelength is

$$\lambda' - \lambda = 0.00006 \text{ nm} = \frac{h}{mc}(1 - \cos\theta) = 0.00243 \text{ nm} (1 - \cos\theta),$$

$$1 - \cos\theta = 0.0247,$$

$$\cos\theta = 1 - 0.0247 = 0.975, \qquad \theta = \boxed{12.7°}.$$

28.9 From the de Broglie relation we have

$$\lambda = \frac{h}{p} = \frac{h}{mv} = \frac{6.63 \times 10^{-34}\ \text{J·s}}{(0.145\ \text{kg})(50\ \text{m/s})} = \boxed{9.1 \times 10^{-35}\ \text{m}}.$$

28.11 From the de Broglie relation we have $\lambda = \frac{h}{p} = \frac{h}{mv}$,

$$v = \frac{h}{m\lambda} = \frac{6.63 \times 10^{-34}\ \text{J·s}}{(9.11 \times 10^{-31}\ \text{kg})(6.0 \times 10^{-9}\ \text{m})} = \boxed{1.2 \times 10^{5}\ \text{m/s}}.$$

28.13 At 1 keV we can assume a nonrelativistic electron with momentum given by $p = \sqrt{2mE}$. The de Broglie relation then gives

$$\lambda = \frac{h}{p} = \frac{h}{\sqrt{2mE}} = \frac{6.63 \times 10^{-34}\ \text{J·s}}{\sqrt{2(9.11 \times 10^{-31}\ \text{kg})(1000\ \text{eV})(1.60 \times 10^{-19}\ \text{J/eV})}},$$

$$\lambda = \boxed{3.88 \times 10^{-11}\ \text{m}}.$$

28.15 First compute the orbital speed of the earth.

$v = \frac{2\pi r}{T}$ where r is the earth radius and T is the period of rotation.

$$v = \frac{2\pi(1.5 \times 10^{11}\ \text{m})}{3.16 \times 10^{7}\ \text{s}} = 2.99 \times 10^{4}\ \text{m/s}.$$

$$\lambda = \frac{h}{mv} = \frac{6.63 \times 10^{-34}\ \text{J·s}}{(5.98 \times 10^{24}\ \text{kg})(2.99 \times 10^{4}\ \text{m/s})} = \boxed{3.7 \times 10^{-63}\ \text{m}}.$$

28.17 We must use relativistic mechanics to solve this problem.

$$E^2 = [KE + m_oc^2]^2 = p^2c^2 + (m_oc^2)^2.$$

$$KE = 100\ \text{MeV} = 1.60 \times 10^{-11}\ \text{J},$$

$$m_oc^2 = (1.67 \times 10^{-27}\ \text{kg})(3.00 \times 10^{8}\ \text{m/s})^2 = 1.50 \times 10^{-10}\ \text{J}.$$

$$p = \frac{1}{c}\sqrt{(KE + m_oc^2)^2 - (m_oc^2)^2} = 2.37 \times 10^{-19}\ \text{kg·m/s}.$$

$$\lambda = \frac{h}{p} = \frac{6.63 \times 10^{-34}\ \text{J·s}}{2.37 \times 10^{-19}\ \text{kg·m/s}} = \boxed{2.8 \times 10^{-15}\ \text{m}}.$$

28.19 (a) First find the speed v. $v = \frac{2\pi r}{T} = \frac{2\pi(0.75\ \text{m})}{1.0\ \text{s}} = 4.71\ \text{m/s}.$

$$\lambda = \frac{h}{mv} = \frac{6.63 \times 10^{-34}\ \text{J·s}}{(0.25\ \text{kg})(4.71\ \text{m/s})} = 5.63 \times 10^{-34}\ \text{m}.$$

The number of wavelengths in a circumference is $n = \frac{2\pi r}{\lambda} = \boxed{8.4 \times 10^{33}}$.

(b) In order for n to be 1000, h must be many times larger.

$$\frac{h_{new}}{h} = \frac{8.4 \times 10^{33}}{1000} = \boxed{8.4 \times 10^{30}}.$$

28.21 First we need to calculate the wavelength of the electrons.

$$\lambda = \sqrt{\frac{1.50}{54}}\ \text{nm} = 0.167\ \text{nm}.$$

The Bragg equation is $n\lambda = 2d \sin\theta$. The angle between incident and diffracted beams is $2\theta = 50°$, so $\theta = 25°$. For first order diffraction n = 1.

$$d = \frac{\lambda}{2 \sin\theta} = \frac{0.167\ \text{nm}}{2 \sin 25°} = 0.197\ \text{nm} = \boxed{0.20\ \text{nm}}.$$

28.23 From the Heisenberg uncertainty relation we get

$\Delta x\, \Delta p = \Delta x\, m\Delta v \geq h/2\pi$.

$$\Delta v \geq \frac{h/2\pi}{m\,\Delta x} = \frac{6.63 \times 10^{-34}\ \text{J·s}/2\pi}{(9.11 \times 10^{-31}\ \text{kg})(0.50 \times 10^{-3}\ \text{m})} = \boxed{0.23\ \text{m/s}}.$$

28.25 The uncertainty in energy is $\Delta E = 10^{-11}$ E, where E is 1.44×10^3 eV. The uncertainty in time is given by

$$\Delta t = \frac{h/2\pi}{\Delta E} = \frac{6.63 \times 10^{-34}\ \text{J·s}/2\pi}{(10^{-11})(14.4 \times 10^3\ \text{eV})(1.6 \times 10^{-19}\ \text{J/eV})} = \boxed{4.58\ \text{ns}}.$$

28.27 We can treat 10.0 MeV protons nonrelativistically. The momentum is then

$$p = \sqrt{2mE} = \sqrt{(2)(1.67 \times 10^{-27}\ \text{kg})(1.60 \times 10^{-12}\ \text{J})}\ ,$$

$p = 7.31 \times 10^{-20}$ kg·m/s.

$\Delta p = 0.0100\ p = 7.31 \times 10^{-22}$ kg·m/s.

$$\Delta x = \frac{h/2\pi}{\Delta p} = \frac{6.63 \times 10^{-34}\ \text{J·s}/2\pi}{7.31 \times 10^{-22}\ \text{kg·m/s}} = \boxed{1.44 \times 10^{-13}\ \text{m}}.$$

28.29 From Problem 28.28 we have

$$\Delta p = \frac{h\,\Delta\lambda}{\lambda^2} = \frac{(6.63 \times 10^{-34}\ \text{J·s})(1.00 \times 10^{-12}\ \text{m})}{(5.00 \times 10^{-11}\ \text{m})^2}$$

$$\Delta p = \boxed{2.65 \times 10^{-25}\ \text{kg·m/s}}.$$

28.31 We are given that the uncertainty in energy over the energy is $\frac{\Delta E}{E} = \frac{1}{10^7}$. For a photon we know that $E = cp$ and so $\Delta E = c\Delta p$.

$$\frac{\Delta E}{E} = \frac{\Delta p}{p} = \frac{(h/2\pi\Delta x)}{p} = \frac{\lambda}{2\pi\,\Delta x}.$$

$$\Delta x = \left(\frac{\lambda}{2\pi}\right)\left(\frac{E}{\Delta E}\right) = \frac{\lambda}{2\pi} \times 10^7.$$

(a) For the x ray: $\Delta x = 10^7\ (0.071\ \text{nm})/2\pi = \boxed{0.11\ \text{mm}}$.

(b) For the light: $\Delta x = 10^7(550\text{ nm})/2\pi = \boxed{0.88\text{ m}}$.

(c) For the radio wave: $\Delta x = 10^7(300\text{ m})/2\pi = \boxed{4.78 \times 10^8\text{ m}}$.

28.33 Start with $\Delta x \Delta p \geq h/2\pi$. For a particle moving in a circle its uncertainty in position along the circle is $\Delta x = r\Delta\theta$. It has angular momentum $L = rp$ and uncertainty in angular momentum $\Delta L = r\Delta p$.

$\Delta x \Delta p = (r\Delta\theta)(\Delta L/r) = \Delta\theta\Delta L \geq h/2\pi$.

28.35 For the one-dimensional square potential well

$$E_n = \frac{h^2}{8mL^2}n^2, \text{ for } n = 1, 2, 3, \ldots$$

$$E_n = \frac{(6.63 \times 10^{-34}\text{ J}\cdot\text{s})^2}{8(9.11 \times 10^{-31}\text{ kg})(0.14 \times 10^{-9}\text{ m})^2}n^2 = 3.08 \times 10^{-18}\, n^2\text{ J}.$$

$E_1 = \boxed{3.1 \times 10^{-18}\text{ J}}$, $E_2 = \boxed{1.2 \times 10^{-17}\text{ J}}$, and $E_3 = \boxed{2.8 \times 10^{-17}\text{ J}}$.

28.37 For the one-dimensional square potential well

$E_n = \dfrac{h^2}{8mL^2}n^2$, for $n = 1, 2, 3, \ldots$ The distance L is found from

$$L = \sqrt{\frac{h^2}{8mE_1}} = \sqrt{\frac{(6.63 \times 10^{-34}\text{ J}\cdot\text{s})^2}{8(9.11 \times 10^{-31}\text{ kg})(6.03 \times 10^{-18}\text{ J})}},$$

$L = \boxed{1.0 \times 10^{-10}\text{ m}}$.

28.39 Assume that the wave function has the form $\psi(x) = \psi_o e^{-\alpha x}$. Then

$$\frac{\psi(1/\alpha)}{\psi(10/\alpha)} = \frac{e^{-1}}{e^{-10}} = e^9 = \boxed{8100}.$$

28.41 Probability $= P \propto e^{-2\alpha\Delta x} = 0.0010$.

Probability of penetrating $2\Delta x$ is

$P \propto e^{-4\alpha\Delta x} = \left(e^{-2\alpha\Delta x}\right)^2 = (0.0010)^2 = \boxed{1.0 \times 10^{-6}}$.

28.43 The L shell corresponds to $n = 2$. There can be $\boxed{\text{2 subshells}}$.

28.45 An f level corresponds to $l = 3$. The number of magnetic substates is $2l + 1 = 2(3) + 1 = \boxed{7}$.

28.47 For the two dimensional potential box we get

$$E = E_x + E_y = \frac{h^2}{8mL^2}n_x^2 + \frac{h^2}{8m(2L)^2}n_y^2 = \frac{h^2}{8mL^2}\left(n_x^2 + \frac{1}{4}n_y^2\right).$$

$\boxed{n_x = 1, n_y = 1, E = \frac{1.25\ h^2}{8mL^2}}$, $\boxed{n_x = 1, n_y = 2, E = \frac{2\ h^2}{8mL^2}}$,

$\boxed{n_x = 1, n_y = 3, E = \frac{3.25\ h^2}{8mL^2}}$, $\boxed{n_x = 1, n_y = 4, E = \frac{5\ h^2}{8mL^2}}$,

$\boxed{n_x = 2, n_y = 1, E = \frac{4.25\ h^2}{8mL^2}}$, $\boxed{n_x = 2, n_y = 3, E = \frac{6.25\ h^2}{8mL^2}}$.

Note, $n_x = 2$ and $n_y = 2$ gives the same energy as $n_x = 1$ and $n_y = 4$.

28.49 From the definition of the Bohr magneton we have

$$\mu_B = \frac{eh}{4\pi m} = \frac{(1.6022 \times 10^{-19}\ C)(6.6261 \times 10^{-34}\ J \cdot s)}{4\pi(9.1094 \times 10^{-31}\ kg)},$$

$$\mu_B = \boxed{9.274 \times 10^{-24}\ J/T}.$$

28.51 For a d level, $m_l = 0, \pm 1, \pm 2$.

$$E_n = m_l \mu_B B.$$

$$E_{+2} - E_{-2} = 4\mu_B B = 4(9.27 \times 10^{-24}\ J/T)(0.10\ T) = \boxed{3.7 \times 10^{-24}\ J}.$$

28.53 (a) $\Delta E = 2\mu_B B = \frac{2\ (9.274 \times 10^{-24}\ J/T)(0.80\ T)}{1.60 \times 10^{-19}\ J/eV} = \boxed{9.28 \times 10^{-5}\ eV}$.

(b) $\lambda = \frac{1240\ eV \cdot nm}{9.28 \times 10^{-5}\ eV} = \boxed{1.34\ cm}$.

28.55 Two 1s states, two 2s states, and one 2p state. In this notation the first number is the value of n and the letter labels the value of l.

28.57 For the N shell, n = 4. The possible values of l are 0, 1, 2, and 3. Each of these subshells has 2(2l + 1) states or $2 + 6 + 10 + 14 = 32$ total states.

28.59 (a) Maximum change occurs for $\theta = 180°$.

$$(\lambda - \lambda')_{max} = \frac{h}{mc}(1 - \cos 180°) = 2\frac{h}{mc} = 2\ (2.426 \times 10^{-12}\ m),$$

$$(\lambda - \lambda')_{max} = \boxed{4.492 \times 10^{-12}\ m}.$$

(b) $\lambda_{(50\ keV)} = \frac{1240\ eV \cdot nm}{50000\ eV} = 0.0248\ nm = 2.48 \times 10^{-11}\ m.$

$$\frac{\Delta\lambda_{max}}{\lambda_{(50\ keV)}} = \frac{4.492 \times 10^{-12}\ m}{2.48 \times 10^{-11}\ m} = \boxed{0.18}.$$

28.61 From the Bragg equation $n\lambda = 2d \sin\theta$.

When $\lambda << d$, then $\sin\theta \approx \theta$. The scattering angle is $2\theta \approx \frac{D/2}{L}$.

For n = 1, we have,

$$\lambda = 2d\,\theta = d\,\frac{D/2}{L}, \quad \text{so that } D = \frac{2L\lambda}{d}.$$

28.63 We want to find $\frac{\Delta f}{f}$. From the Planck relation we get

$$\frac{\Delta f}{f} = \frac{\Delta E}{E} \approx \frac{h/2\pi\Delta t}{E} = \frac{(4.14 \times 10^{-15}\ \text{eV·s})/2\pi(2.0 \times 10^{-8}\ \text{s})}{2.4\ \text{eV}} = \boxed{1.4 \times 10^{-8}}.$$

28.65 $\Delta E = 2\{\mu_B B\} = \frac{hc}{\lambda}$.

$$B = \frac{hc}{2\lambda\mu_B} = \frac{(6.63 \times 10^{-34}\ \text{J·s})(3.00 \times 10^{8}\ \text{m/s})}{2(2.00 \times 10^{-2}\ \text{m})(9.27 \times 10^{-24}\ \text{J/T})} = \boxed{0.536\ \text{T}}.$$

28.67 The wave function must satisfy boundary conditions in three dimensions.

$\psi(x,y,z) = \psi_o \sin(n_x\pi x/L_x) \sin(n_y\pi y/L_y) \sin(n_z\pi z/L_z)$.

The corresponding energy comes from computing

$$\frac{\Delta(\Delta\psi/\Delta x)}{\Delta x} + \frac{\Delta(\Delta\psi/\Delta y)}{\Delta y} + \frac{\Delta(\Delta\psi/\Delta z)}{\Delta z},$$

$$E = \frac{h^2}{8m}\left[\left(\frac{n_x}{L_x}\right)^2 + \left(\frac{n_y}{L_y}\right)^2 + \left(\frac{n_z}{L_z}\right)^2\right].$$

Chapter 29

THE NUCLEUS

29.1 Using Eq.(29.1) with $m_U = m_H$, we find

$$\frac{v'_H}{v'_N} = \frac{m_H + 14m_H}{m_H + m_H} = \frac{15}{2} = \boxed{7.5}.$$

29.3 Using Eq.(29.1) with $m_U = m_H$ and with $24m_H$ substituted for the nitrogen, we find

$$\frac{v'_H}{v'_{Mg}} = \frac{m_H + 24m_H}{m_H + m_H} = \frac{25}{2} = \boxed{12.5}.$$

29.5 We are given that $A = 29$. We are also given that $N = A - Z = Z + 1$. So $Z = (A - 1)/2 = 28/2 = 14$.

The element with $Z = 14$ is silicon. The isotope is $\boxed{{}^{29}_{14}\text{Si}}$.

29.7 ${}^{39}_{19}\text{K}$ has $\boxed{19 \text{ protons}}$ and $39 - 19 = \boxed{20 \text{ neutrons}}$.

${}^{39}_{20}\text{Ca}$ has $\boxed{20 \text{ protons}}$ and $39 - 20 = \boxed{19 \text{ neutrons}}$.

${}^{23}_{11}\text{Na}$ has $\boxed{11 \text{ protons}}$ and $23 - 11 = \boxed{12 \text{ neutrons}}$.

${}^{23}_{12}\text{Mg}$ has $\boxed{12 \text{ protons}}$ and $23 - 12 = \boxed{11 \text{ neutrons}}$.

29.9 ${}^{200}_{80}\text{Hg}$ has $\boxed{80 \text{ protons}}$ and $200 - 80 = \boxed{120 \text{ neutrons}}$.

${}^{16}_{8}\text{O}$ has $\boxed{8 \text{ protons}}$ and $16 - 8 = \boxed{8 \text{ neutrons}}$.

${}^{232}_{90}\text{Th}$ has $\boxed{90 \text{ protons}}$ and $232 - 90 = \boxed{142 \text{ neutrons}}$.

29.11 The average atomic mass is computed from the relative abundances as
$M = 24 \times 0.787 + 25 \times 0.101 + 26 \times 0.112 = \boxed{24.3}$.

29.13 The radius of the nucleus depends on the number of nucleons through
$r = R_0 A^{1/3} = (1.2 \times 10^{-15} \text{ m})(235)^{1/3} = \boxed{7.4 \times 10^{-15} \text{ m}}$.

29.15 The radius of the nucleus depends on the number of nucleons through
$r = R_0 A^{1/3}$. Thus the ratio of radia of two nuclei is given by

$$\frac{r_{Th}}{r_{Fe}} = \left(\frac{A_{Th}}{A_{Fe}}\right)^{1/3} = \left(\frac{232}{57}\right)^{1/3} = \boxed{1.6}.$$

29.17 The Einstein relationship for mass and energy is $E = mc^2$. For mass of 1 u,

$E = (1.66054 \times 10^{-27}\ \text{kg})(2.99792 \times 10^{8}\ \text{m/s}^2)^2 = 1.49241 \times 10^{-10}\ \text{J}$.

$$E = \frac{1.49241 \times 10^{-10}\ \text{J}}{1.6022\ \text{J/eV}} = 9.315 \times 10^{8}\ \text{eV} = \boxed{931.5\ \text{MeV}}.$$

29.19 From Fig. 29.4:

Binding energy of ${}^{58}_{28}\text{Ni} \approx 58$ nucleons (8.8 MeV/nucleon) $= \boxed{510\ \text{MeV}}$.

Binding energy of ${}^{232}_{90}\text{Th} \approx 232$ nucleons (7.5 MeV/nucleon) $= \boxed{1740\ \text{MeV}}$.

29.21 For ${}^{9}_{4}\text{Be}$:

$$\begin{aligned} 5m_n &= 5(939.566\ \text{MeV}) = 4697.830\ \text{MeV} \\ 4m_H &= 4(938.783\ \text{MeV}) = \underline{3755.132\ \text{MeV}} \\ & \qquad\qquad 8452.962\ \text{MeV} \\ \text{Less mass of } {}^{5}_{2}\text{He} & \qquad\qquad \underline{-8394.796}\ \text{MeV} \\ \text{Binding energy} &= \boxed{58.166\ \text{MeV}}. \end{aligned}$$

29.23 (a) The gravitational force is given by $F_G = \frac{Gm^2}{r^2}$ and the electrostatic force (the Coulomb force) is $F_E = \frac{ke^2}{r^2}$.

$$\frac{F_G}{F_E} = \frac{Gm^2}{ke^2} = \frac{(6.67 \times 10^{-11}\ \text{N}\cdot\text{m}^2/\text{kg}^2)(1.67 \times 10^{-27}\ \text{kg})^2}{(8.99 \times 10^{9}\ \text{N}\cdot\text{m}^2/\text{C}^2)(1.60 \times 10^{-19}\ \text{C})^2},$$

$$\frac{F_G}{F_E} = \boxed{8.08 \times 10^{-37}}.$$

(b) $\boxed{\text{No}}$. Gravity is not important in nuclear structure.

29.25 ${}^{234}_{92}\text{U} \rightarrow {}^{4}_{2}\text{He} + {}^{230}_{90}\text{X}$.

The daughter nucleus is $\boxed{{}^{230}_{90}\text{Th}}$.

29.27 ${}_{84}\text{Po} \rightarrow {}^{4}_{2}\text{He} + {}_{82}\text{X}$.

The resulting element is $\boxed{\text{lead}}$.

29.29 $\text{X} \rightarrow {}^{0}_{-1}\beta + {}^{90}_{39}\text{Y}$,

$\text{X} = \boxed{{}^{90}_{38}\text{Sr}}$.

29.31 (a) $^{226}_{88}\text{Ra} \rightarrow ^{4}_{2}\text{He} + ^{222}_{86}\text{X}$.

The daughter nucleus is $\boxed{^{222}_{86}\text{Rn}}$.

(b) $^{233}_{91}\text{Pa} \rightarrow ^{0}_{-1}\beta + ^{233}_{92}\text{X}$.

The daughter nucleus is $\boxed{^{233}_{92}\text{U}}$.

(c) $^{59}_{26}\text{Fe} \rightarrow ^{0}_{0}\gamma + ^{59}_{26}\text{X}$.

The daughter nucleus is $\boxed{^{59}_{26}\text{Fe}}$.

29.33 (a) First we need to compute the Q value.

$$Q = [m(^{238}_{92}\text{U}) - m(^{4}_{2}\text{He}) - m(^{234}_{90}\text{Th})]c^2,$$

$$Q = [238.05078 \text{ u} - 4.002603 \text{ u} - 234.04359 \text{ u}]c^2.$$

Total KE released $= Q = 0.00459 \text{ u} \times 931.5 \text{ MeV/u} = \boxed{4.27 \text{ MeV}}$.

29.35 $^{234}_{92}\text{U} \rightarrow \boxed{^{230}_{90}\text{Th} \rightarrow ^{226}_{88}\text{Ra} \rightarrow ^{222}_{86}\text{Rn} \rightarrow ^{218}_{84}\text{Po} \rightarrow ^{214}_{82}\text{Pb}}$.

29.37 (a) $\dfrac{234}{4} = 58 + \dfrac{2}{4}$.

$^{234}_{91}\text{Pa}$ belongs to the series with $A = 4n + 2$, the $\boxed{\text{Uranium series}}$.

(b) $\dfrac{231}{4} = 57 + \dfrac{3}{4}$.

$^{231}_{91}\text{Pa}$ belongs to the series with $A = 4n + 3$, the $\boxed{\text{Actinium series}}$.

29.39 $^{3}_{1}\text{H} \rightarrow ^{3}_{2}\text{He} + \beta^-$.

Conservation of energy requires that

$$Q = \left[m_N(^{3}_{1}\text{H}) - m_N(^{3}_{2}\text{He}) - m_e\right]c^2, \text{ where } m_N \text{ is the nuclear mass.}$$

If we add and subtract one electron mass, we get

$$Q = \left[m_N(^{3}_{1}\text{H}) + 1m_e - m_N(^{3}_{2}\text{He}) - m_e - m_e\right]c^2,$$

$$Q = \left[m_a(^{3}_{1}\text{H}) - m_a(^{3}_{2}\text{He})\right]c^2, \text{ where } m_a \text{ is the atomic mass.}$$

$Q = [m(^3_1H) - m(^3_2He)]c^2 = 2809.433 \text{ MeV} - 2809.414 \text{ MeV},$

$Q = \boxed{0.019 \text{ MeV}}.$

29.41 The reaction is $^A_ZX \rightarrow {}^A_{Z+1}Y + {}^0_{-1}\beta.$

Conservation of energy requires that

$$Q = \left[m_N(^A_ZX) - m_N(^A_{Z+1}Y) - m_e \right] c^2.$$

If we add and subtract Z electron masses, we get

$$Q = \left[m_N(^A_ZX) + Zm_e - m_N(^A_{Z+1}Y) - Zm_e - m_e \right] c^2,$$

$$Q = \left[m_a(^A_ZX) - m_a(^A_{Z+1}Y) \right] c^2.$$

Thus spontaneous emission can occur when $m_a(^A_ZX) > m_a(^A_{Z+1}Y)$.

29.43 (a) First we need to compute the Q value.

$Q = [m(^{235}_{92}U) - m(^4_2He) - m(^{231}_{90}Th)]c^2,$

$Q = [235.04392 \text{ u} - 4.002603 \text{ u} - 231.03630 \text{ u}]c^2.$

Total KE released $= Q = 0.00502 \text{ u} \times 931.5 \text{ MeV/u} = \boxed{4.67 \text{ MeV}}.$

(b) Kinetic energy available to the alpha particle is

$$KE_\alpha \approx \frac{A}{A+4} Q = \frac{231}{235} 4.67 \text{ MeV} = \boxed{4.59 \text{ MeV}}.$$

29.45 The momentum of the daughter is m_dv_d and that of alpha is $m_\alpha v_\alpha$. Conservation of momentum requires that $m_\alpha v_\alpha + m_dv_d = 0.$

The total KE = Q, so that $\frac{1}{2}m_\alpha v_\alpha^2 + \frac{1}{2}m_dv_d^2 = Q.$

Eliminating v_d from the equation for Q gives

$$\frac{1}{2}m_\alpha v_\alpha^2 + \frac{1}{2}\frac{1}{m_d}m_\alpha^2 v_\alpha^2 = \frac{1}{2}m_\alpha v_\alpha^2(1 + m_\alpha/m_d) = Q,$$

$$KE_\alpha = \frac{Q}{1 + m_\alpha/m_d}.$$

29.47 (a) $D(\text{in Gy}) = \dfrac{H(\text{in Sv})}{Q} = \dfrac{5.0 \text{ Sv}}{1} = \boxed{5.0 \text{ Gy}}.$

(b) $D(\text{in Gy}) = \dfrac{H(\text{in Sv})}{Q} = \dfrac{5.0 \text{ Sv}}{10} = \boxed{0.50 \text{ Gy}}.$

29.49 $^{13}_{6}C + ^{1}_{0}n \rightarrow ^{14}_{6}X + ^{0}_{0}\gamma$. X is $\boxed{^{14}_{6}C}$.

29.51 $^{10}_{5}B + ^{4}_{2}He \rightarrow \boxed{^{14}_{7}N}$.

29.53 $KE = kT = (1.38 \times 10^{-23}\ J/K)(300\ K)\left(\frac{1}{1.6 \times 10^{-19}\ J/eV}\right) = 2.59 \times 10^{-2}\ eV$,

$KE = 2.59 \times 10^{-2}\ eV\left(\frac{1\ MeV}{10^6\ eV}\right) = \boxed{2.59 \times 10^{-8}\ MeV}$.

29.55 $E = \left[2m(^{2}_{1}H) - m(^{1}_{0}n) - m(^{3}_{2}He)\right]c^2$,

$E = 2(1876.125\ MeV) - 939.566\ MeV - 2809.414\ MeV = \boxed{3.27\ MeV}$.

29.57 We use the relationship that $r = R_o A^{1/3}$ and that $V = \frac{3}{4}\pi r^3$.

$$A = \frac{r^3}{R_o^3} = \frac{3V}{4\pi R_o^3} = \frac{3(1.51 \times 10^{-42}\ m^3)}{4\pi(1.2 \times 10^{-15}\ m)^3} = 209.$$

If the nucleus has N = 126 neutrons, then its charge is
$Z = A - N = 209 - 126 = 83$.

The element is $\boxed{^{209}_{83}Bi}$.

29.59 Total KE = Q, momentum of daughter is $m_d v_d$ and that of alpha is $m_\alpha v_\alpha$.
Conservation of momentum requires $m_\alpha v_\alpha + m_d v_d = 0$.

Energy conservation gives $\frac{1}{2} m_\alpha v_\alpha^2 + \frac{1}{2} m_d v_d^2 = Q$.

Use the momentum equation to eliminate v_α from the energy equation.

$$\frac{1}{2}\frac{1}{m_\alpha} m_d^2 v_d^2 + \frac{1}{2} m_d v_d^2 = \frac{1}{2} m_d v_d^2 (m_d/m_\alpha + 1) = Q,$$

$$KE_d = \frac{Q}{1 + m_d/m_\alpha} \approx \frac{Q}{1 + A/4} = \boxed{\frac{4}{A+4}Q}.$$

29.61 (a) The initial kinetic energy must be great enough to overcome the repulsive potential. Set $KE = PE = k\frac{q^2}{r} = (9 \times 10^9\ N{\cdot}m^2/C^2)\frac{3.2 \times 10^{-19}\ C}{10^{-14}\ m}$,

$KE = \boxed{9.2 \times 10^{-14}\ J = 0.58\ MeV}$.

(b) Using $E = k_BT$, we get

$$T = \frac{E}{k_B} = \frac{9.2 \times 10^{-14}\ \text{J}}{1.38 \times 10^{23}\ \text{J/K}} = \boxed{6.6 \times 10^9\ \text{K}}.$$

29.63 Find the activity in Bq: $33000\ \text{Ci} \times 3.7 \times 10^{10}\ \text{Bq/Ci} = 1.22 \times 10^{15}\ \text{Bq}$.

The disintegration constant λ can be found from the half life.

$$\lambda = \frac{0.693}{t_{1/2}} = \frac{0.693}{(12.3\ \text{y})(365\ \text{d/y})(24\ \text{h/d})(3600\ \text{s/h})} = 1.79 \times 10^{-9}\ \text{s}^{-1}.$$

Now we can use the relationship that $A = \lambda N$ to find the number of atoms from the ratio of the activity to the disintegration constant.

$$N = \frac{A}{\lambda} = \frac{1.22 \times 10^{15}\ \text{s}^{-1}}{1.79 \times 10^{-9}\ \text{s}^{-1}} = 6.82 \times 10^{23}\ \text{nuclei}.$$

6.02×10^{23} atoms of tritium $= 3$ g.

$$6.82 \times 10^{23}\ \text{atoms}\ \frac{3\ \text{g}}{6.02 \times 10^{23}\ \text{atoms}} = \boxed{3.4\ \text{g}}.$$

29.65 In an elastic collision we conserve momentum and kinetic energy. Use the very good approximation that $m_\alpha = 4m_p$.

Momentum: $m_\alpha v_\alpha = m_p v_p + m_\alpha v_\alpha'$,

KE: $m_\alpha v_\alpha^2 = m_p v_p^2 + m_\alpha v_\alpha'^2$.

Use the momentum equation to eliminate v_α'. Then

$m_\alpha v_\alpha^2 = m_p v_p^2 + [m_\alpha v_\alpha - m_p v_p]^2/m_\alpha$,

$m_\alpha v_\alpha^2 = m_p v_p^2 + m_\alpha v_\alpha^2 - 2m_p v_\alpha v_p + m_p^2 v_p^2/m_\alpha$,

$4v_\alpha^2 = v_p^2 + 4v_\alpha^2 - 2v_p v_\alpha + v_p^2/4$,

$0 = v_p - 2v_\alpha + v_p/4$,

$v_p + v_p/4 = \frac{5}{4} v_p = 2\,v_\alpha$,

$v_p = \frac{8}{5} v_\alpha = \boxed{1.6\ v_\alpha}$.

Chapter 30

LASERS, HOLOGRAPHY, AND COLOR

30.1 The length L is the number of half wavelengths times half a wavelength.

$$L = n\frac{\lambda}{2}. \quad n = \frac{2L}{\lambda} = \frac{2 \times 0.300\text{ m}}{632.8 \times 10^{-9}\text{ m}} = \boxed{9.48 \times 10^5}.$$

30.3 The separation between modes was derived in the text and given as Eq. (30.2).

$$\Delta f = \frac{c}{2L} = \frac{3.00 \times 10^8\text{ m/s}}{2 \times 0.355\text{ m}} = \boxed{423\text{ MHz}}.$$

30.5 $$E = \frac{hc}{\lambda} = \frac{1240\text{ eV}\cdot\text{nm}}{694.3\text{ nm}} = \boxed{1.786\text{ eV}}.$$

30.7 $$E = \frac{hc}{\lambda} = \frac{1240\text{ eV}\cdot\text{nm}}{543\text{ nm}} = \boxed{2.28\text{ eV}}. \quad \boxed{\text{Green.}}$$

30.9 From the definition of visibility given in Eq. (30.3) we have

$$\frac{I_{max} - I_{min}}{I_{max} + I_{min}} = 0.40.$$

$$I_{max} - I_{min} = 0.40\, I_{max} + 0.40\, I_{min},$$

$$0.60\, I_{max} = 1.40\, I_{min}, \qquad \frac{I_{max}}{I_{min}} = \frac{1.40}{0.60} = \boxed{2.3}.$$

30.11 From the definition of visibility given in Eq. (30.3) and from knowledge that for even illumination $\gamma = V$, we have

$$\gamma = V = \frac{I_{max} - I_{min}}{I_{max} + I_{min}} = \frac{10 - 1}{10 + 1} = \frac{11}{9} = \boxed{0.82}.$$

30.13 From the definition of visibility given in Eq. (30.3) and the definition of coherence given in Eq. (30.4) we have

$$\gamma = \frac{I_1 + I_2}{2\sqrt{I_1 I_2}} \frac{I_{max} - I_{min}}{I_{max} + I_{min}} = \frac{3 + 1}{2\sqrt{3}} \frac{(5.5 - 1)I_{min}}{(5.5 + 1)I_{min}} = \boxed{0.80}.$$

30.15 From the definition of visibility given in Eq. (30.3) and the definition of coherence given in Eq. (30.4) we have

$$V = \frac{2\sqrt{I_1 I_2}}{I_1 + I_2}\gamma = \frac{2\sqrt{1.5}}{2.5}\, 0.80 = \boxed{0.78}.$$

$(I_{max} + I_{min})V = I_{max} - I_{min}$, so $I_{max}(1 - V) = I_{min}(1 + V)$.

$$\frac{I_{max}}{I_{min}} = \frac{1 + V}{1 - V} = \frac{1.78}{1 - 0.78} = \boxed{8.1}.$$

30.17 The spatial frequency is given by $F = \sin\theta/\lambda$.

$\sin\theta = \lambda F = 488 \times 10^{-9}\text{ m} \times 10^3\text{ mm/m} \times 1500\text{ lines/mm} = 0.732.$

$\theta =$ $\boxed{47.1°}$. The field of view is limited by a cone making an angle of 47.1° from the direction of the reference beam.

30.19 The spatial frequency is given by $F = 1/d = \sin\theta_1/\lambda_1$.

$$\sin\theta_2 = \frac{\lambda_2}{d} = \frac{\lambda_2}{\lambda_1}\sin\theta_1 = \frac{546.1}{632.8}\sin 28°.$$

$\theta_2 = 23.9°$. Three beams at $\boxed{0° \text{ and } \pm 23.9°}$.

30.21 Use the grating equation: $n\lambda = d\sin\theta$, where $n = 1$.

$$\theta_1 = \sin^{-1}\frac{\lambda_1}{d} = \sin^{-1}(589.6\times10^{-6}\ \text{mm}\times1500/\text{mm}) = 62.18°.$$

$$\theta_1 = \sin^{-1}\frac{\lambda_2}{d} = \sin^{-1}(589.0\times10^{-6}\ \text{mm}\times1500/\text{mm}) = 62.07°.$$

$\Delta\theta = \theta_1 - \theta_2 = \boxed{0.11°}$.

30.23 Use the Wien displacement law: $\lambda T = 2.90\times10^{-3}$ m·K.

$$T = \frac{2.90\times10^{-3}\ \text{m·K}}{967\times10^{-9}\ \text{m}} = \boxed{3000\ \text{K}}.$$

30.25 To achieve the same illumination $P_1\times0.20 = P_{light} = P_2\times0.035$.

$$\frac{P_2}{P_1} = \frac{0.20}{0.035} = \boxed{5.7}.$$

30.27 (a) $\boxed{\text{Orange}}$, (b) $\boxed{\text{purple}}$.

30.29 (a) $\boxed{\text{Green}}$, (b) $\boxed{\text{green}}$, (c) $\boxed{\text{red}}$, (d) $\boxed{\text{no light emerges}}$.

30.31 410.2 nm, $\boxed{\text{violet}}$; 434.0 nm, $\boxed{\text{violet}}$; 486.1 nm, $\boxed{\text{blue}}$; 656.3 nm, $\boxed{\text{red}}$.

30.33 In the small angle approximation for the double slit equation, $x = \frac{L\lambda}{D}$. The frequency and wavelength are related through $c = f\lambda$.

$$\Delta\lambda = \lambda_1 - \lambda_2 = c\left(\frac{1}{f_1} - \frac{1}{f_2}\right) \approx c\frac{\Delta f}{f^2} = \Delta f\frac{\lambda^2}{c}.$$

$$\Delta x = \frac{L\,\Delta\lambda}{d} = \frac{L}{d}\Delta f\frac{\lambda^2}{c} = \frac{2500\ \text{mm}}{0.20\ \text{mm}}(2\times10^9/\text{s})\frac{(633\times10^{-9}\ \text{m})^2}{3.0\times10^8\ \text{m/s}} = 3.3\times10^{-8}\ \text{m},$$

$\Delta x = \boxed{3.3\times10^{-5}\ \text{mm}}$.

Chapter 31

CONDENSED MATTER

31.1 In the CsCl crystal, each atom of one type, say Cs, is surrounded by a cubic array of atoms of the other type, say Cl. Thus theNearest neighbor distance is one half length of the body diagonal. The body diagonal is a vector that is one unit of a in the x direction, one unit in the y direction, and one unit in the z direction. Here, a is the interatomic distance along the cube edge. The body diagonal has a length $l = \sqrt{a^2 + a^2 + a^2} = \sqrt{3}\, a.$

$$d_{nn} = \frac{1}{2}\sqrt{3}\, a = \frac{\sqrt{3}}{2} 0.411 \text{ nm} = \boxed{0.356 \text{ nm}}.$$

31.3 Refer to the drawing of the CsCl crystal structure in the text. The spacing between nearest neighbor Cs ions is the same as the edge length, $\boxed{0.411 \text{ nm}}$.

31.5 From the diagram of the diamond crystal in the text, we see that the nearest neighbor distance is diamond is one-fourth the body diagonal. The corner to corner distance along the body diagonal is $\sqrt{3}\, a_o$. (See the solution to Problem 31.1.) The distance between nearest neighbors is $\boxed{\frac{\sqrt{3}}{4} a_o}$. Here a_o is the edge length of the cubic unit cell.

31.7 The fermi energy is $E_F = 4.7$ eV.

$$T_F = \frac{E_F}{k} = \frac{4.7 \text{ eV}}{8.62 \times 10^{-5} \text{ eV/K}} = \boxed{5.5 \times 10^4 \text{ K}}.$$

31.9 The density of electrons is N/V where

$$\frac{N}{V} = \frac{\rho \times N_A \times 1 \text{ electrons/atom}}{\text{atomic mass}}$$

$$\frac{N}{V} = \frac{0.82 \text{ g/cm}^3 \times 6.02 \times 10^{23}\text{electrons/mol}}{39.1 \text{ g/mol}}$$

$$\frac{N}{V} = 1.26 \times 10^{22} \text{ electrons/cm}^3 = \boxed{1.26 \times 10^{28} \text{ electrons/m}^3}.$$

31.11 (a) $E = E_F + \phi = (3.10 + 2.74) \text{ eV} = \boxed{5.84 \text{ eV}}$.

(b) $\lambda = \frac{hc}{E} = \frac{4.41 \times 10^{-15} \text{ eV·s} \times 3.00 \times 10^8 \text{ m/s}}{5.84 \text{ eV}} = \boxed{212 \text{ nm}}$.

31.13 From the definition of mobility we get

$v = \mu E = (1.47 \times 10^{-3} \text{ m}^2\text{/V·s})(10^{-3} \text{ V/cm} \times 100 \text{ cm/m}),$

$v = \boxed{1.47 \times 10^{-4} \text{ m/s}}$.

31.15 Start with the relationship between mobility and collision time,

$\mu = \frac{e\tau}{m}$. Rearrange to get

$$\tau = \frac{m\mu}{e} = \frac{9.11 \times 10^{-31}\ \text{kg} \times 1.47 \times 10^{-3}\ \text{m}^2/\text{V}\cdot\text{s}}{1.60 \times 10^{-19}\ \text{C}} = \boxed{8.37 \times 10^{-15}\ \text{s}}.$$

31.17 The conductivity σ is the reciprocal of the resistivity ρ. The conductivity is related to the mobility by

$\sigma = \frac{1}{\rho} = ne\mu$. Upon rearranging we find

$$\mu = \frac{1}{\rho ne} = \frac{1}{(1.59 \times 10^{-8}\ \Omega\cdot\text{m})(5.85 \times 10^{28}/\text{m})(1.60 \times 10^{-19}\ \text{C})},$$

$$\mu = \boxed{6.72 \times 10^{-3}\ \text{m}^2/\text{V}\cdot\text{s}}.$$

31.19 (a) $\mu = \frac{\sigma}{ne} = \frac{6.17 \times 10^7/\Omega\cdot\text{m}}{5.76 \times 10^{28}/\text{m}^3 \times 1.60 \times 10^{-19}\ \text{C}} = \boxed{6.69 \times 10^{-3}\ \text{m}^2/\text{V}\cdot\text{s}}$.

(b) $\tau = \frac{m\mu}{e} = \boxed{3.81 \times 10^{-14}\ \text{s}}$.

31.21 $\lambda = \frac{hc}{E} = \frac{4.14 \times 10^{-15}\ \text{eV}\cdot\text{s} \times 3.00 \times 10^8\ \text{m/s}}{0.67\ \text{eV}} = \boxed{1850\ \text{nm}}$.

The radiation is in the $\boxed{\text{infrared}}$.

31.23 $\lambda = \frac{hc}{E} = \frac{4.14 \times 10^{-15}\ \text{eV}\cdot\text{s} \times 3.00 \times 10^8\ \text{m/s}}{1.14\ \text{eV}} = \boxed{1090\ \text{nm}}$.

31.25 The conductivity includes contributions from both the motion of electrons and the motion of the holes.

$$\sigma = ne(\mu_n + \mu_p) = 9.1 \times 10^{12}/\text{m}^3 \times 1.60 \times 10^{-19}\ \text{C}\ (8600 + 250)10^{-4}\ \text{m}^2/\text{V}\cdot\text{s},$$

$$\sigma = \boxed{1.29 \times 10^{-6}/\Omega\cdot\text{m}}.$$

31.27 Positive charges move in the direction of the current density, the positive x direction. The effect of the magnetic field B_z deflects the charges in the - y direction, thus setting up an electric field pointing along the + y direction. Negative charges move opposite to the direction of the current density. They are deflected by the magnetic field into the - y direction, setting up an electric field that points along the - y direction.

31.29 The Hall field is given by

$$E_y = \frac{j_x B_z}{ne} = \frac{(0.125\ \text{A})(1.0\ \text{T})}{(2 \times 5 \times 10^{-6}\ \text{m}^2)(4.6 \times 10^{22}/\text{m}^3)(1.6 \times 10^{-19}\ \text{C})},$$

$E_y = \boxed{1.7 \text{ V/m}}$.

The Hall voltage is $V_y = E_y l_y = (1.7 \text{ V/m})(5 \times 10^{-3} \text{ m}) = \boxed{8.5 \text{ mV}}$.

31.31 The conductivity includes contributions from both the motion of electrons and the motion of the holes.

$\sigma = e[n\mu_n + p\mu_p]$,

$\sigma = e\left[(5.1 \times 10^{18}/\text{m}^3)(0.86 \text{ m}^2/\text{V·s}) + (1.6 \times 10^{7}/\text{m}^3)(0.0250 \text{ m}^2/\text{V·s})\right]$,

$\sigma = \boxed{0.70/\Omega\text{·m}}$.

31.33 The junction potential is given by

$$V_o = -\frac{kT}{e}\ln\frac{n_p}{n_n} = -(8.62 \times 10^{-5} \text{ eV·K}^{-1}/\text{e})(300 \text{ K}) \ln\frac{1.0 \times 10^{5}}{5.0 \times 10^{14}} = \boxed{0.58 \text{ V}}.$$

31.35 From the rectifier equation we get

$I = I_o\left(e^{eV/kT} - 1\right)$.

The combination kT/e has the value of 0.0259 eV, so

$I = 0.130\ \mu\text{A}\left(e^{-1.40/0.0259} - 1\right) = \boxed{-0.130\ \mu\text{A}}$.

31.37 (a) We showed that for a full rectified wave the average value of the voltage is related to the peak value of the voltage through

$\overline{V} = \frac{2V_o}{\pi}$, where the peak value $V_o = \sqrt{2}\, V_{rms}$.

$$\overline{V} = \frac{2\sqrt{2}}{\pi} V_{rms} = \frac{2\sqrt{2}}{\pi}(15 \text{ V}) = \boxed{13.5 \text{ V}}.$$

(b) $V_{rms} = \boxed{15 \text{ V}}$.

31.39 (a) First find the average voltage. $\overline{V} = \frac{2V_o}{\pi} = \frac{2(9.3 \text{ V})}{\pi} = 5.9 \text{ V}$.

The average current is found from

$$\overline{I} = \frac{\overline{V}}{R} = \frac{5.9 \text{ V}}{470\ \Omega} = \boxed{13 \text{ mA}}.$$

(b) The rms current is found from

$$I_{rms} = \frac{V_{rms}}{R} = \frac{V_o/\sqrt{2}}{R} = \frac{9.3 \text{ V}}{\sqrt{2}\, 470\ \Omega} = \boxed{14 \text{ mA}}.$$

(c) $P = I_{rms}V_{rms} = (14 \text{ mA})(9.3 \text{ V})/\sqrt{2} = \boxed{92 \text{ mW}}$.

31.41 The wavelength is determined from the band gap energy.

$$\lambda = \frac{hc}{E} = \frac{(4.14 \times 10^{-15} \text{ eV·s})(3.00 \times 10^{8} \text{ m/s})}{1.4 \text{ eV}} = \boxed{890 \text{ nm}}.$$

31.43 (a) flux × area × efficiency = power available.

$$\text{area} = \frac{40 \text{ kW}}{(1.0 \text{ kW/m}^2)(0.10)} = \boxed{400 \text{ m}^2}.$$

(b) flux × area × efficiency × cos θ = power available.

$$\text{area} = \frac{400 \text{ m}^2}{\cos 45°} = \boxed{570 \text{ m}^2}.$$

31.45 (a) From Eq. (31.2) we have $E_F = \frac{h^2}{2m}\left(\frac{3}{8\pi}\frac{N}{V}\right)^{2/3}$,

$$E_F = \frac{(6.63 \times 10^{-34} \text{ J·s})^2}{2(9.11 \times 10^{-31} \text{ kg})}\left(\frac{3}{8\pi} 1.08 \times 10^{28}/\text{m}^3\right)^{2/3},$$

$$E_F = 2.86 \times 10^{-19} \text{ J} = \boxed{1.79 \text{ eV}}.$$

(b) $T_F = \frac{E_F}{k} = \frac{1.79 \text{ eV}}{8.62 \times 10^{-5} \text{ eV·K}^{-1}} = \boxed{2.08 \times 10^4 \text{ K}}.$

31.47 (a) When the end of the transformer connected to the diode is negative relative to the end connected to point B, the diode will conduct so that point A is negative relative to B. When the transformer voltage is reversed, the diode will block so that the potential from A to B is zero. Thus the wave form is that of a half rectified negative sine wave.

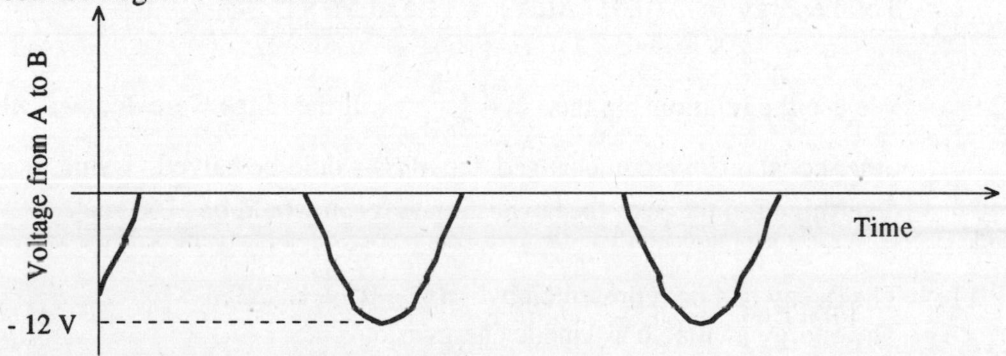

(b) The signal is that of a half rectified sine wave. The average voltage is one half that of the average voltage of $2V_o/\pi$ of a full rectified sine wave. In this case the voltage from A to B is negative, so the average voltage is

$$\overline{v} = -V_o/\pi = (-12.0 \text{ V})/\pi = \boxed{3.82 \text{ V}}.$$

31.49 $I = I_o\left(e^{eV/kT} - 1\right)$. $\frac{I + I_o}{I_o} = e^{eV/kT}$.

$$V = \frac{kT}{e}\ln\left(\frac{I + I_o}{I_o}\right) = 2.59 \times 10^{-2} \text{ V} \ln\left(\frac{10 \times 10^{-3} + 0.10 \times 10^{-6}}{0.10 \times 10^{-6}}\right),$$

$$V = \boxed{0.30 \text{ V}}.$$

Chapter 32

ELEMENTARY PARTICLE PHYSICS

32.1 Use Einstein's equation: $E = mc^2$.

$$m = \frac{E}{c^2} = \frac{1.000 \text{ MeV}}{(2.9979 \times 10^8 \text{ m/s})^2}(1.6022 \times 10^{-13} \text{ J/MeV})$$

$$m = \boxed{1.783 \times 10^{-30} \text{ kg}}.$$

32.3 The energy required is twice the rest mass energy of one of the particles, say the electron. $E = 2(m_ec^2) = 2(0.511 \text{ MeV}) = \boxed{1.022 \text{ MeV}}$.

32.5 Use the Einstein equation $E = mc^2$ for each person-antiperson.

$E = 2(mc^2) = 2(60 \text{ kg})(2.998 \times 10^8 \text{ m/s})^2 = \boxed{1.08 \times 10^{19} \text{ J}}$.

The Hoover dam produces about 2.0×10^{16} J/y.

$$\text{Time} = \frac{1.08 \times 10^{19} \text{ J}}{2.0 \times 10^{16} \text{ J/y}} = \boxed{540 \text{ years}}.$$

32.7 Total energy $= KE_{total} + 2(m_ec^2) = 2\left(\frac{1}{2}m_ec^2\right) + 2(m_ec^2) = 3(m_ec^2)$,

Total energy $= 3(0.511 \text{ MeV}) = \boxed{1.533 \text{ MeV}}$.

32.9 We use the relationship that $m \approx \dfrac{h}{2\pi R_oc}$. If the range were doubled while all other constants were unchanged, the mass would be halved. Using the value determined in the text, the new estimate for m would be $\boxed{160 \ m_e}$.

32.11 The decay can be represented by $\pi^+ \rightarrow \mu^+ + \nu$.
The energy available as kinetic energy is,

$KE = (m_\pi - m_\mu)c^2 = 139.6 \text{ MeV} - 105.7 \text{ MeV} = \boxed{33.9 \text{ MeV}}$.

Note that we have treated the mass of the neutrino as zero. Even if it is nonzero, it is so small compared to the mass of the μ that we can ignore it.

32.13 (a) For the K^o the distance traveled is the speed times the time: $d = vt$. Here the time is the half life. We need to use the relativistic formula for the time as measured in the laboratory. Thus the distance becomes

$$d = vt = \frac{vt_o}{\sqrt{1 - v^2/c^2}} = \frac{0.90(3.00 \times 10^8 \text{ m/s})(0.89 \times 10^{-10} \text{ s})}{\sqrt{1 - (0.90)^2}},$$

$$d = \boxed{5.5 \text{ cm}}.$$

(b) For the $K^\pm$: $d = \dfrac{0.90(3.00 \times 10^8 \text{ m/s})(1.24 \times 10^{-8} \text{ s})}{\sqrt{1 - (0.90)^2}} = \boxed{7.68 \text{ m}}$.

32.15 From the de Broglie relationship we get

$$p = \frac{h}{\lambda} \approx \frac{6.63 \times 10^{-34}\ \text{J·s}}{10 \times 10^{-15}\ \text{m}} \approx \boxed{7 \times 10^{-20}\ \text{kg·m·s}^{-1}}.$$

32.17 $KE = \frac{1}{2}mv^2 = mgh = (0.25\ \text{kg})(9.8\ \text{m/s}^2)(1.0\ \text{m})\left(\frac{1\ \text{eV}}{1.6 \times 10^{-19}\ \text{J}}\right)\left(\frac{1\ \text{TeV}}{10^{12}\ \text{eV}}\right)$,

$KE = \boxed{1.5 \times 10^{7}\ \text{TeV}}$.

32.19 From the de Broglie relationship we get

$$\lambda = \frac{h}{p} = \frac{ch}{E} = \frac{(3.00 \times 10^{8}\ \text{m/s})(6.63 \times 10^{-34}\ \text{J·s})}{(10 \times 10^{9}\ \text{eV})(1.60 \times 10^{-19}\ \text{J/eV})} = \boxed{1.24 \times 10^{-16}\ \text{m}}.$$

32.21 (a) Recall that the ideal gas law can be expressed in the form PV = NkT, where N is the number of molecules and k is the Boltzmann constant.

$$\frac{N}{V} = \frac{P}{kT} = \frac{1.5 \times 10^{-6}\ \text{Pa}}{(1.38 \times 10^{-23}\ \text{J/K})(288\ \text{K})} = \boxed{3.8 \times 10^{14}/\text{m}^3}.$$

(b) The barometric formula is $P(z) = P_o e^{-mgz/kT}$. The mass of an oxygen molecule is the molecular mass divided by Avogadro's number.

$$m = \frac{\text{molecular mass}}{N_A} = \frac{32 \times 10^{-3}\ \text{kg/mol}}{6.02 \times 10^{23}/\text{mol}} = 5.32 \times 10^{-26}\ \text{kg}.$$

$$z = \frac{kT}{mg} \ln \frac{P_o}{P(z)} = \frac{(1.38 \times 10^{-23}\ \text{J/K})(288\ \text{K})}{(5.32 \times 10^{-26}\ \text{kg})(9.8\ \text{m/s}^2)} \ln\left(\frac{1.0 \times 10^{5}\ \text{Pa}}{1.5 \times 10^{-6}\ \text{Pa}}\right),$$

$z = \boxed{190\ \text{km}}$.

32.23 (b) $\boxed{\text{Does not conserve charge}}$.

(c) $\boxed{\text{Does not conserve baryon number}}$.

(d) $\boxed{\text{Does not conserve baryon number or charge}}$.

32.25 The composition of the proton is given in Table 32.4 as uud. The antiparticle is the $\bar{p}$, so its composition is $\boxed{\bar{u}\bar{u}\bar{d}}$.

32.27 (a) For the udd: charge = 0 and baryon number = 1. $\boxed{\text{Can occur}}$.

(b) For the $\bar{u}$dd: charge = $-\frac{1}{3}$ and baryon number = $\frac{1}{3}$. $\boxed{\text{Can not occur}}$.

(c) For the dsu: charge = 0 and baryon number = 1. $\boxed{\text{Can occur}}$.

(d) For the d$\bar{u}$: charge = - 1 and baryon number = 0. $\boxed{\text{Can occur}}$.

32.29 $\Delta m = m_{initial} - m_{final} = m(\pi^-) + m(p) - m(\Lambda^o) - m(K^o)$,

$\Delta m = [139.6 + 938.3 - 1116 - 497.7]\ \text{MeV} = \boxed{-536\ \text{MeV}}$.

The reaction requires energy in order to take place.

32.31 (a) Each water molecule is composed of one oxygen atom and two hydrogen atoms for a total of 10 protons per molecule. The number of molecules containing 2.5×10^{32} protons is:

$$\text{number} = \frac{2.5 \times 10^{32}\ \text{protons}}{10\ \text{protons/molecule}} = \boxed{2.5 \times 10^{31}\ \text{molecules}}.$$

(b) $$\text{Mass} = \frac{\text{number of molecules} \times \text{mass/mol}}{\text{molecules/mol}},$$

$$\text{mass} = \frac{2.5 \times 10^{31} \times 0.018\ \text{kg/mol}}{6.02 \times 10^{23}/\text{mol}} = \boxed{7.5 \times 10^{5}\ \text{kg}}.$$

$$\text{Volume} = \frac{\text{mass}}{\text{density}} = \frac{7.5 \times 10^{5}\ \text{kg}}{1000\ \text{kg/m}^3} = \boxed{750\ \text{m}^3}.$$